21st Century Feature Writing

21st Century Feature Writing

Carla Johnson

Saint Mary's College
Notre Dame, Indiana

PEARSON

and

Boston   New York   San Francisco
Mexico City   Montreal   Toronto   London   Madrid   Munich   Paris
Hong Kong   Singapore   Tokyo   Cape Town   Sydney

Executive Editor: *Karon Bowers*
Series Editor: *Molly Taylor*
Editorial Assistant: *Michael Kish*
Senior Marketing Manager: *Mandee Eckersley*
Composition and Prepress Buyer: *Linda Cox*
Manufacturing Buyer: *JoAnne Sweeney*
Cover Administrator: *Kristina Mose-Libon*
Editorial-Production Service: *Matrix Productions Inc.*
Illustrations: *Modern Graphics, Inc.*
Electronic Composition: *Modern Graphics, Inc.*

For related titles and support materials, visit our online catalog at www.ablongman.com.

Between the time website information is gathered and then published, it is not unusual for some sites to have closed. Also, the transcription of URLs can result in unintended typographical errors. The publisher would appreciate notification where these errors occur so that they may be corrected in subsequent editions.

Library of Congress Cataloging-in-Publication Data

Johnson, Carla.
    21st century feature writing / Carla Johnson.
        p. cm.
    Includes bibliographical references and index.
ISBN 0-205-38015-8
    1. Feature writing. I. Title: Twenty-first century feature writing. II. Title.
PN4784.F37J65 2004
070.4'4—dc22

2003063930

Printed in the United States of America

10 9 8 7 6 5 4 3 2 1   09 08 07 06 05 04

This book is dedicated to my son, Aaron H. Hoffman,
my first student, who has carried the burden of a writer's child
to early heights in his own career.

Contents

Preface   xiii

Acknowledgments   xv

PART ONE • *The Feature Story*   1

1   *Dimensions of Feature Writing*   3

What Makes a Feature Different?   4

Emotion, Human Interest, and Personal Style   5
  Click Here: It's the Way You See It   6

Sidebar: Model Feature for Writers   7

"Flight in Bomber Evokes Wartime Memories," Andrea Dominello   8
  Chat Room   10
  Create   10

"Making Stone Soup," Molly Donnellon   10
  Chat Room   13
  Create   13

Special Section: The Professional Writing Portfolio   14

"The Internship Payoff," Nellie Williams   15
  View   16
  Help   17

2   *Feature Writing as a Process*   18

Idea Generation   19

Research   21

Organization   21

Context   23

Drafting and Revision   24
  Click Here: Outline for the Donahue Story   26

"Troy Donahue," Carla Johnson   27
   Chat Room   29
   Create   29

"Men's Pickup Lines, Then and Now," Renée Donovan and Sara Pendley   30
   Chat Room   31
   Create   31

"I Am an American: Ken Takaki and World War II," Evelyn Gonzales   32
   View   34
   Help   34

3   *The Style and Structure of the Feature Story*   36

Feature Structure   37

The Lead: How Do I Begin?   38

How Do I Continue?   41

How Do I Stop?   43

Publication Style   44

Personal Style   44
   Click Here: Stylus—Engraving Your Mark   46

"Colm Feore," Carla Johnson   46
   Chat Room   49
   Create   49

"After the Cheering Stops," Eric Hansen   50
   Chat Room   52
   Create   52

"A Moveable Moscow: A Good Hotel Near Red Square," Jennifer Kaye Jones   53
   Chat Room   54
   Create   55
   View   55
   Help   56

4   *Research*   58

Public Records   59

Library Resources   59

The Internet   61

Personal Observation   62

Click Here: Those Useful Citations, Max J. Skidmore   64

"Injustice Undone: Part 1: The Crimes," Matthew S. Galbraith   66
Chat Room   68
Create   69

"Roxie Hart," Carla Johnson   69

"Sean Savage," Holly M. James   71
Chat Room   73
Create   74
View   74
Help   75

5   *Interviews*   76

Types of Interviews   76

Preparation for the Interview   78

The Time Factor   80

Conducting the Interview   82

Ending the Interview   83
Click Here: A Question of Ethics   84

"James Earl Jones," Carla Johnson   85
Chat Room   88
Create   88

"Working Man: Tom Wopat," Carla Johnson   89
Chat Room   91
Create   92

"Something Old, Something New," Kate Dooley   92
Create   94
View   94
Help   95

6   *Attitude and Interpretation: Details   96*

Attitude and Tone   97

Interpretation   97

Organization of Material   100

Quotations   100
Click Here: About Attitude and Tone   101

Travel Features   101

"Marin Mazzie: Living Her Dream," Carla Johnson   102

" 'Rent' Reclaims Broadway for U.S.," Carla Johnson   104

"Sweet Onion Charm," Holly M. James   105

Sidebar: "Vidalia Onion Festival: April 10–14, 2003," Amelia Michalski   108
> Chat Room   109
> Create   109

"Chicago: City of Cheap Thrills," Emily Ford   109

"Two Beaches," Tara Blanchard   112

> Chat Room   114
> Create   114
> View   114
> Help   115

PART TWO • *Feature Formats*   117

7   *The Magazine Industry*   119

The Magazine   120

The Magazine Feature   121

The Magazine Feature Market   122
> Click Here: Crafting a Query Letter, Kristina V. Jonusas   123

Sidebar: Resources for Magazine Writers   124

"Sports Massage Is His Passion," Susanne M. Alexander   125
> Chat Room   128
> Create   128

"A Writer's Words," Ann Basinski   128

"Disaster Search and Rescue Dogs," Kristin Mehus-Roe   132
> Chat Room   135
> Create   135
> View   136
> Help   136

8   *The Newspaper Industry*   137

The Newspaper   138

The Newspaper Feature   139

Click Here: Check *Facts* or *Fail* 142

Sidebar: Resources for Newspaper Writers **142**

"Free Falling," Denise McGuire, Katie Miller **144**

"Dining High," Christina Reitano, Holly M. James **144**

"The Newest Member of the Footwear Family," Jennika Kirkbride **144**

"Marz Sweet Shop," Carla Johnson **146**
    Chat Room 150
    Create 150

"A Slice of Comfort," Molly V. Strzelecki **150**
    Chat Room 151
    Create 152
    View 152
    Help 153

9 **The Internet** **154**

The Internet Industry **155**

Internet Story Structure: The Package **155**

Linear vs. Nonlinear Writing **156**

Writing the Internet Feature **157**

The Roundup: An Experiment **158**

The Writer Online **158**

"No. 1 Reason to Visit South Bend—The Cake Nazi," Mary Ellen Brown **160**

"No. 2 Reason to Visit South Bend—There's No Right Way," Sara Pendley **161**

"No. 10 Reason to Visit South Bend—You Can Fiddle Away at Fiddler's," Renée Donovan **162**
    Chat Room 163
    Create 164

"Smiling in Terror: An Interview with MC Paul Barman," Nick Roskelly **164**
    Chat Room 166
    Create 166
    View 166
    Help 167

10 **The Public Relations Feature** **168**

The Purpose of Public Relations Writing **169**

Public Relations Proposals and Research   *170*

Types of Public Relations Features   *170*
  Click Here: Public Relations Speechwriting   171
  Click Here: Direction Sheet, Kaitlin E. Duda   172

The Release Format   *175*

"Tune In to Reality," Kaitlin E. Duda   *174*
  Chat Room   176
  Create   177

"John of the Archives," Sarah K. Magness   *177*
  Chat Room   178
  Create   179
  View   179
  Help   180

11 *Special Feature Formats*   *181*

Opinion Pieces   *182*

Reviews   *183*

"8:16 a.m.," Megan Colvin   *184*
  Chat Room   186
  Create   186
  Click Here: More on Those Pesky Pronouns, Nick Roskelly   187

"A Leader Much Like Ourselves," Carla Johnson   *188*

"Looking to Get Hitched, Just Not Yet" Molly V. Strzelecki   *189*

"Judeo Jerry," Carla Johnson   *191*
  Chat Room   191
  Create   192

"Some Like It Hot," Carla Johnson   *192*

"Chicago," Marianne Orfanos, Laura Coristin   *193*

"Common's Electric Circus (MCA)," Nick Roskelly   *194*
  Chat Room   195
  Create   195
  View   196
  Help   196

Bibliography   **197**

Index   **199**

Preface

Before I could even write, I was a storyteller. As a toddler, I drew narrative cartoons and sewed their pages sequentially inside cloth covers to make picture books. I grew up in a small town and, when reading and writing entered my vocabulary, avidly read the newspaper that came from the nearby "big city" of South Bend, Ind. The writers whose names appeared in bylines intrigued me as wondrous creatures who provided daily stories to multitudes of people. Half a century later, I have a body of published works in a variety of genres—poetry, short stories, public relations releases, academic articles and a book on global advertising. However, my favorite genre has been journalism, writing feature stories for newspapers and magazines, most of them for The South Bend Tribune. I recognized long ago that the mass media writer reaches everyone, not just an elite audience of poets or academics, and that was what I wanted to do.

This book comes from a lifelong passion for storytelling and a commitment to write stories and teach others how to write stories that are at once entertaining and instructive and ethically sound.

I looked at the stack of feature writing textbooks that are on the market and recognized that I had something unique to contribute. This book, "21st Century Feature Writing," draws on 25 years of practical experience as a journalist; it also draws on two decades of teaching experience. I'd initially envisioned compiling a reader of my favorite celebrity profiles, but my editor at Allyn & Bacon had other ideas. Challenged to write a full-blown feature writing textbook, I knew exactly where to turn. Students learn as much from reading other students' writing as they do from reading the masterpieces of Pulitzer Prize winners. As I've taught feature writing using drafts of this book, my students have contributed their writing as well as their feedback on the chapters. I've changed what didn't work and published what did.

I also bring to this book the perspective of someone who entered the field in the midst of a major revolution in journalism. My training and experience straddle pre- and post-Watergate journalism as well as the movement known as "New Journalism." I grew up while Dwight D. Eisenhower occupied the presidency, and I really don't think I'd have wanted to be a journalist if The South Bend Tribune had run stories about his affair with his World War II driver. My sense of what is in the public interest and what constitutes fair play derives from what I suppose would be called the "old school." The style in which I wrote my stories for The South Bend Tribune resembles the Washington Post's style: third-person voice, clean and objective. During my 25 years as a special assignment writer for The Tribune, I was not allowed to use first person in my feature stories even when I reported on events in which I participated. I learned to convey my opinion in reviews without ever referring to myself in any way, even obliquely as "the writer." Nevertheless, I admire the work of journalists who write more personally, such as Dominick Dunne, Calvin Trillin and Anna Quindlen. The Tribune's writers of the new generation have become decidedly present in their stories, so the revolution continues. I've started to make concessions for my students' use of subjective voice, and this book reflects that flexibility.

Now I watch to see what long-term effects the 9-11 tragedy may have on American journalism.

However we investigate our stories, cast them into a style, and justify their content, the heart and core of feature writing remains unchanged. This book is for those who love language, have a compelling desire to tell stories and are curious about people, about politics, about places—about life. You may have guessed by now that I'm not a kid anymore, but I still wake up each day with a sense of wonder about our world and a drive to gain insight into its mysteries, which—when I find them—I am glad to report to anyone willing to indulge me.

I offer this book to students of all ages who share a spirit of discovery. Storytelling dates back to the beginning of time, and it will go on till the end. Isocrates once said that human expertise with language distinguishes us from the wild beasts and paves the way for democracy: Through language "we have come together and founded cities and made laws and invented arts." It was the democratic nature of journalism that drew me to it in the first place. My goal has always been, in some small way, to enhance the lives and elevate the thoughts of the common, everyday folks who read newspapers and magazines.

Journalism may not generally be viewed as a high calling, but a good writer can make it one.

Dr. Carla Johnson
Notre Dame, Indiana
12 July 2003

Acknowledgments

The author wishes to acknowledge her family's love, support and understanding: Aaron H. Hoffman, Helen Jackson, Marie Remington, James R. Jackson and, of course, Sunny. I also acknowledge with deepest gratitude the contributions of other individuals whose assistance and contributions have been invaluable: my Saint Mary's College student interns Megan Colvin, Jennika Kirkbride, Allison Scarnecchia, Katie-Nell Scanlon, Arianna Stella and especially Kristina V. Jonusas, who wrote a number of chapter Views, Chat Rooms and Create activities; my Feature Writing teaching assistants Kaitlin Duda, Elisa Ryan and Renée Donovan; Mary Beth Dominello, Celia Fallon and Tod A. Moorhead at Saint Mary's College; and Susanne M. Alexander, Andrea Dominello, Matthew S. Galbraith, Eric Hansen, Kristin Mehus-Roe, Amelia Michalski, Nick Roskelly, Max J. Skidmore and Molly V. Strzelecki, who so freely shared their professional talents.

I also would like to extend my gratitude to the manuscript reviewers: Carolyn M. Byerly, Ithaca College; David Byland, Oklahoma Baptist University; Kay Colley, University of North Texas; Julia R. Fox, Northern Illinois University; Elizabeth Fraas, Eastern Kentucky University; Loren C. Gruber, Missouri Valley College; Tina M. Harris, University of Georgia; Michael B. Hesse, University of South Alabama; Thom Lieb, Towson University; Chris Ogbondah, University of Northern Iowa; Rosalie S. Peck, Concord College; Larry Timbs, Winthrop University; David Wendelken, James Madison University; and Steve Wiegenstein, Culver-Stockton College. Thanks also to Carol Schaal for her collaboration on an early proposal.

Additionally, I want to thank the many editors for whom I've worked at The South Bend Tribune for the opportunities and guidance they offered over three decades—in particular, Chuck Small, who was a student in my journalism class at St. Joseph's High School long before he became my editor at The Tribune. Most of all, of course, I must acknowledge all my students, past and present, who not only contributed to the book but also were continuous reviewers of it. As they read their assignments in early versions of this textbook, students endlessly caught mistakes and gave crucial feedback on what they liked and did not like. Beyond that, they showed immense generosity in allowing me to publish their stories here. Their struggles and victories flow through these pages.

21st Century Feature Writing

Part I

The Feature Story

1

Dimensions of Feature Writing

I went to Kansas City and worked on the Star. It was regular newspaper work: Who shot whom? Who broke into what? Where? When? How? But never why. Not really why.
—Ernest Hemingway

Staff writer Bill Moor's story on the front page of The South Bend Tribune on Dec. 23, 1999, covered developments in a scandal involving a local police officer. The headline, "Feirrell loses police power," signaled that this was a breaking news story. You might expect the lead for such a story to be written in inverted pyramid style, answering the questions posed by the five Ws to which Hemingway refers in the opening quote (*who, what, where, when, why + how*). However, Moor's suspended-interest lead has the characteristics of a feature lead:

> Charlie Feirrell is no longer being paid by taxpayers.
> He no longer has police powers.
> And the soap opera in which the St. Joseph County police officer is the main character doesn't seem to be going away.

The lead withholds key facts to create suspense. It also expresses the writer's personal view, or interpretation, of unfolding events: "the soap opera . . . doesn't seem to be going away." Calling a news event a "soap opera" is, in itself, the sort of personal filtering that a decade ago would have been viewed as inappropriate for hard news. That the scandal "doesn't seem to be going away" calls for speculation, opinion. He calls the officer by his nickname, "Charlie," instead of his given name, "Charles," which had been used in previous news stories.

Only in the fourth paragraph does Moor's copy begin to sound like news: "On Wednesday, Sheriff Rick Seniff announced that he and Feirrell had reached an agreement in which Feirrell gave up his duties as a police officer while also drawing a paycheck from his accrued vacation as well as sick and compensation days until Feb. 5, 2000."

The rest of the story presents unfiltered information until, midway, the writer once again asserts his presence: "So just before Seniff and the county Police Department's attorney, Steve Hostetler, held a news conference, the five Merit Board members put on their winter coats and left the county police road patrol office." Traditionally, the objective nature of news reporting has required writers to function outside the story. It has been the special domain of the feature writer to get inside the story, to observe and interpret detail. Had Moor written just the facts, he would have stated that "the five Merit Board members left the county police road patrol office before the news conference." The previous paragraph gives the reason for the walkout: an agreement on a Merit Board issue had been reached between the sheriff and the attorney before the board had been able to meet and have a voice in the decision. Moor's description of the board members donning their coats and going out into the winter cold dramatizes the event and colors it emotionally—you'd have to be really upset to walk out on a meeting you'd had to bundle up and drive through the snow to attend. The image set by the weather also suggests that the board had been "shut out in the cold," that is, excluded from the decision.

What Makes a Feature Different?

The stories we see on the front pages of newspapers are typically *hard news* stories, written in *inverted pyramid structure*. The most crucial information must be given in the first paragraph, the *lead*, with subsequent paragraphs arranged in descending importance of the information disclosed in them. The final paragraphs should be dispensable, meaning the reader wouldn't miss key facts if those paragraphs were eliminated, which they sometimes will be in the process of page layout to make copy fit into available space.

Throughout the 1990s, front-page newspaper stories, which for decades had been considered hard news, were going "soft." Feature leads and obvious writer opinion blurred the line that previously separated sections of the newspaper. What appears on the front page of the paper in the 21st century may bear stylistic resemblance to stories that appear in such sections as entertainment and travel.

Media critics and readers alike have criticized mainstream journalism for its sensationalized news reporting, previously the domain only of the tabloids. Studies conducted by the American Society of Newspapers Editors and the Radio and Television News Directors Association following the death of John F. Kennedy Jr., showed that the American public believed the media overcovered and miscovered the news, especially when it came to celebrities.[1] The terrorist attacks on Sept. 11, 2001, changed the media agenda. The dominant story in the weeks preceding the attacks was Congressman Gary Condit's relationship with missing intern Chandra Levy. News commentators quickly admitted that this story now seemed trivial and inappropriate; Condit next made the news when his name was announced for an appointment to a domestic security committee. A year later, Jim Naughton, president of the Poynter Institute for Media Studies, described the new climate as "sobered, focused, and concerned."[2]

Nevertheless, fierce competition among the media, including the rise of tabloid journalism and threats posed by newsmagazines, television and the Internet, may have blurred lines between news and feature writing forever. As those who deliver the news fought the war for ratings and subscriptions, they looked to the feature for ways to entice viewers and readers, often turning to the feature's storytelling structure to appeal to readers, "relating events chronologically, spinning out a narrative that ends in a satisfying conclusion."[3] The casting of news into feature structure has led to a new story category, the *news feature*. Some news stories are said to have been given *feature treatment*. For decades the top-rated television show, "60 Minutes," has offered the electronic version of the news feature.

So what differentiates the feature story from the news story today? As Moor's story demonstrates, differences have become less pronounced. It's also true that news and feature writers go through similar processes as they work to produce their final product. As Chip Scanlan of the Poynter Institute has said, "Good writing may be magical, but it's not magic."[4] Whether it's "a deadline account of a fatal accident," an editorial or a politician's profile, both news and feature writing require the same process—reporting, researching, focusing, organizing, drafting and rewriting.

Still, in our world of 24-hour news cycles, the feature remains more reflective and often goes more in depth than the news story. The creatively rich, rainbow world of feature writing brings together the skills and talents associated with both the news reporter and the fiction writer because the feature story requires greater emphasis on

- emotion
- human interest
- personal style

Of course, both news and feature writers produce *stories*. A story essentially requires the elements that Hemingway identified: who did what, where, when and why. While these basic components of story have not varied much throughout human history, the purposes of the storytellers have been quite diverse. The purpose of a story may be to inform, to educate, to inspire, to record historical events, to entertain and/or to transmit cultural mores.[5] The storyteller's purpose has dictated the shape, or structure, that the story will take. The feature may fulfill any of these purposes.

Emotion, Human Interest and Personal Style

While the news story may also capture human interest, its requisite objectivity precludes the greater stylistic freedom and the coloration of details with the author's emotional responses of the more subjective feature. The features that follow exemplify these feature story characteristics. Andrea Dominello's "Flight in Bomber Evokes Wartime Memories" displays a huge difference between a news story and a feature—she adds her own emotions and reflects personally on the human-interest

---

🖰 *Click Here: It's the Way You See It*

A new episode of "Ally McBeal," aired on Jan. 10, 2000, interwove three seemingly unrelated stories with a unifying idea—we judge people by outward appearances. This unifying principle allowed "McBeal" creator David E. Kelley to offer a prism of perspectives on a single idea. In other words, he told three stories, each with a different perspective—or angle. The angle reflects the storyteller's point of view.

Unlike the news story, a feature is shaped by the writer's point of view. The way the writer sees a situation, person, place or thing determines the way the story will be presented. Kelley presented the idea of image from three different angles: Ally fell for a homeless man after he showered and put on some new clothes; Billy boosted his masculine image by hiring an entourage of glazed-eyed models to surround him, in a parody of 1980s pop star Robert Palmer, to hook a macho client; and Ling's escort service preyed upon the insecurities of adolescent males who were less than great looking. His presentations evoked viewer emotion and thought. The sad conclusion of Ally's brief love affair tugged at the viewer's emotions; misrepresented and misread appearances created the tragic outcome. Billy's success at snaring the client provoked thought—perhaps Billy's outrageous image truly reflected an aspect of his inner self. The homely teen who used Ling's escort service discovered his escort would have liked him whether he paid for her services or not. The show made viewers ponder when and how much outward appearances should matter.

When you propose a story idea to an editor, don't be surprised if you are asked, "What's your angle?" You are being asked not only about your point of view but also about the story's human-interest appeal. Why would readers want to read this story? The word *angle* entered the English language as the Old English word *angul*, which meant "fishhook" or "hook." In journalism today, the story's angle refers to an approach that will "hook" the reader. Throughout time, storytellers have revealed universal truths about the human condition. A story's angle appeals to our universal human interest in our shared contradictions, complexities, triumphs and tragedies.

---

components of the story she has written. Although she crafts most of the feature in objective third person, at times she refers to herself in order to convey the insights she is experiencing. "I've never been terribly interested in history," she writes. The movie "Pearl Harbor" was "just another love story to me."

Whether the feature writer's personal style is lyrical or technical, humorous or empathetic, that style must serve the story. Andrea's story and Molly Donnellon's first-person account, written as a class assignment, demonstrate how a match of story and style can result in an inspirational story. Molly's story may *seem* to be a story about Molly Donnellon; as you read it, keep in mind that stories written for newspaper and magazine must finally tell about the experiences and lives of others. They must place that information in a universal context to which almost anyone can relate in some way.

Notice how Molly does this.

The two features are *participatory features*. The writers have gone outside their own experience to learn about the experiences of others; they have entered another person's world. The purpose of Andrea's flight with a World War II pilot was to see how it must have felt for him to fly this plane during the war. Molly volunteered to work at an institution where she interacted with people whose daily experience is outside the norm.

The first time I taught this chapter, the one you have just read, the students produced exemplary participatory features. Happy with this success, I gave the same assignment the following semester. It was February and flu season, and you might say that Molly's story became infectious. The Molly Donnellon flu caused everyone to write exactly like her, right down to the italicized stream of consciousness.

From the days of the ancient Greeks, young artists have imitated the masters as a way to explore form and style. For novice writers today, imitation still serves its purpose as a useful exercise—writers tell their own stories within another writer's sentence, paragraph and/or organizational structures. Imitations of Ernest Hemingway work well (see Jennifer Jones' "A Moveable Moscow" in Chapter 3). The first Create activity on p. 13, asks you to imitate Molly's style as an exercise.

That's not what I originally assigned, however. The students were asked to write participatory features modeled after Andrea's bomber story. In the classroom as in the professional world, it's a good idea to be sure that you understand the assignment before you begin to write.

☞ Sidebar: Model Features for Writers

The chapters in this book offer model features written by the author, by college students and by professional writers, including some prize winners. To locate features that have won the Pulitzer Prize, here's a suggested link: http://pbs.org/newshour/essays-dialogues.html. To read features honored by the Society of Professional Journalists, go to the SPJ Web site, www.spj.org. In addition to listing SPJ award winners, the site offers partial and full texts of prize-winning stories and links to the news organizations' own Web sites. Special issues of Quill, the society's magazine, also lists winners and describes their entries.

The American Society of Magazine Editors' Web site, http://asme.magazine.org, lists and describes the annual winners of the National Magazine Awards. Interested readers may locate the magazine issues in which the winning stories appeared and find the text as it was published.

It is crucial for novice feature writers to read, read, read—everything from cereal boxes at breakfast to features published in the campus newspaper to stories published in major newspapers and magazines.

Feature Story

"Flight in Bomber Evokes Wartime Memories"

Andrea Dominello
Author of "Flight in Bomber Evokes Wartime Memories." Photo courtesy of Olan Mills/2003.

Andrea Dominello published the following feature in the Indian Hill Journal, Loveland, Ohio, on July 11, 2001. Gary Presley, the newspaper's executive editor, sent Andrea an e-mail detailing his response to the piece: "It's too long," he said, but "you correctly used a lot of terminology that I'm sure you're not familiar with. That's not easy to do. Your writing flows very well. Your weaving in Pearl Harbor and your thoughts (or lack thereof) regarding war was well done. It will show older readers that younger folks can care about such things, and gives younger readers a portal into the history by allowing them to identify with you. Most people would have tried to just write about 'my flight on a big old airplane.' Your column transcended that and got to the people. You might have evoked a little more of the flight, but you evoked enough to let folks know you were there. Nice job." A day later, she received an e-mail from a 30-year-old reader who commented that too many young people "take America and its liberties for granted."

Jack Keller was 17 when he walked into the kitchen and saw his parents crouched over the radio in the living room crying. It was Dec. 7, 1941.

He was told to hush, and not a word was spoken as they listened to the broadcast informing them that the Japanese attacked Pearl Harbor—and the United States was at war.

"That was the day I knew I wanted to go," said Jack.

Jack's older brother Sam was stationed on the USS Blackfin at Pearl Harbor, and although his parents couldn't stand the thought of losing their sons to war, Jack was determined to go—without hesitation.

He fibbed about his age and tried to join the ranks, but he was told to come back when he turned 18. That he did. He enlisted in 1942 and was assigned to the U.S. Army 8th Air Force, third bomb division, fourth bomb wing, 447th bomb group, 711th bomb squadron stationed in Rattlesden, England.

He was put at the position of "tail gunner," who mans the machine gun at the tail end of the B-17. The 77-year-old North College Hill man explained what it was like to be squeezed into the confined area at the back of the plane on his knees for entire missions—sometimes up to nine hours.

"I wore an electric-heated suit over my long johns and clothes," he said. "It was 50 to 60 degrees below zero."

Jack flew 26 missions, and was injured on only one mission—he still has the souvenir metal fragments in his head to this day. His brother was lucky too, and lived long enough to retire after 30 years with the Navy.

Jack still has the telegram postmarked Dec. 10, 1941, Pearl Harbor, from his brother indicating he was alive and well.

We looked through his memorabilia—flight records and mission details, silver and gold coins from dignitaries including Gen. Henry Shelton, a member of the Joint Chiefs of

Staff; a letter of encouragement from Gen. Dwight Eisenhower sent the night before D-Day, June 6, 1944; and a photo of him and Gen. Paul Tibbets, the pilot of the B-29 bomber "Enola Gay," the plane that dropped the atomic bomb on Hiroshima.

A picture of his original crew reveals a younger Jack, standing tall in front of a B-17 in his brown bomber jacket, which is weathered and worn today.

Jack recalls how his crew named their plane "Ol' Scrapiron" after their first pilot, Lt. Robert Stevenson, who was killed on his first mission en route to Berlin. "Ol' Scrapiron" eventually went down in March 1945 in Holzwickede, Germany, after it collided midair with another B-17 and spiraled to the ground.

I see that reminiscing can be hard—that digging up old memories can sting. Jack said he didn't personally know the men who died on the beaches of Normandy on June 6, 1944. But with tears in his eyes he confesses, "I prayed on my knees that day for those guys."

I've never been terribly interested in history—I saw "Pearl Harbor" and it was just another love story to me. I used to wonder why and how my dad could sit and watch documentaries on the war and Hitler for hours. And when the Persian Gulf War broke out when I was a freshman in high school, I never thought it was a big deal. War was just this far-off thing that didn't affect my world.

History entered my world that day when Jack and I both took our first ride in a restored World War II B-25 bomber called "Old Glory" at Lunken Airport's Million Air center.

The event drew crowds of people, some veterans, and others just with a passion for the planes and their history.

As Jack and I boarded up through the belly of the bomber and put on our earphones to block out the deafening sound of the engines, I wondered how Jack felt 58 years ago when he boarded a B-17 bound for combat. How did he feel looking through the glass window of the tail flying through 88mm German anti-aircraft ammunition called "flak," or seeing other planes being shot out of the sky right in front of him.

"I was scared," he said. "I was so nervous before a mission I would make myself sick. I just prayed we would get back safe.

"I grew up fast. I went from an 18-year-old kid to a 19-year-old man."

Charles Fitzpatrick of Mount Washington said he was considered an old man when he flew at age 26. "I flew B-25s and B-26s from 1941–1945 . . . in the North African invasion. I was forced down three times."

"It's very nostalgic up there," said Bill Smetts of Mason, who also took the ride with us. He flew B-24s in the South Pacific at night looking for Japanese ships trying to supply Chinese troops.

Bill Timmerman of Delhi flew B-25s with the 13th Jungle Air Force in the Pacific theater. "There are a lot of memories up there," he said.

Jack and the other veterans shared their stories of what type of aircraft they flew, where they were stationed, and how many missions they went on. They joked about how they were getting ready for their "final mission" now, as an estimated 1,500 World War II veterans die each day. It gave me chills to see these grown men tearfully embracing each other as if they were old friends.

I feel sorry that I've always taken my freedom and fortuity for granted. I now have a wonderful sense of gratitude—not only to Jack, but to all the veterans who gave themselves to serve our country—without hesitation.

Sixteen million Americans went to war to save the world from totalitarianism—and I thank those people who sacrificed so much—some their own lives, for our nation.

Chat Room

1. Both e-mail responses to Andrea's "Bomber" story comment on a similar theme. What is that theme?
2. The story begins with a flashback lead. Discuss the impact of beginning with the flashback instead of starting with, for example, her ride in the airplane.
3. In our age of 24-hour news cycles, the feature offers the now-rare opportunity for reflection. Identify and discuss the reflective paragraphs in the story.
4. Unlike a news story, a feature may reveal the writer's emotions. Identify Andrea's emotions. Find sentences that explicitly express these emotions.

Create

1. Write your own e-mail response to Andrea Dominello at andrearemke@yahoo .com. She is now editor for three counties at the newspaper's Kentucky office.
2. Rewrite the lead of this feature in inverted pyramid style.
3. Talk to someone who participates in an activity you know little about, such as an athlete, a photographer, a music composer. Write down everything you have learned or better understand about this activity or experience. Then write a participatory feature.
4. Locate the October 2002 issue of Harper's magazine. Read John Jeremiah Sullivan's "Horseman, Pass By," the winner of the 2003 National Magazine Award for feature writing. Make notes on passages that convey emotion and human interest. Describe Sullivan's style.

FEATURE STORY

"Making Stone Soup"

In this personal experience feature, written in 1999 when she was a senior at Saint Mary's College, Molly Donnellon recounts her experience as a volunteer at Portage Manor in South Bend, Indiana. The *personal experience feature* is a participatory story written in the first person. In some cases, writers may perform a role, work or otherwise participate in an activity in order to write about it. In this case, Molly didn't volunteer so that she could write about the mentally ill. Her work was part of the course requirement for People and Justice, a justice education class. Her purpose in writing was to share the insight she gained through her work. Molly's style is as personal as her content. Italicized contrast paragraphs create a sort of dramatic monologue, similar to the asides to the audience—unheard by other characters onstage—employed in theatre to convey what a character is thinking. Although the story has not previously been published, it has become required reading for students enrolled in the course.

What am I doing here? I cannot believe I talked myself into volunteering at this place. What do I know about people who are mentally ill? I know that I have never worked with anyone who has a mental illness—that's what I know.

It is my first day at Portage Manor, a residence home for adults who are mentally ill or handicapped. As I step into the large foyer of the stately mansion, I spot a woman out of the corner of my eye who is holding a book and rocking back and forth, back and forth, while sitting on a small bench.

Oh great. This is going to be just like that movie "One Flew Over the Cuckoo's Nest" with people running around screaming and carrying on, or else just sitting around talking to themselves. All that stuff I told myself about pushing my comfort level and how experiences like this will help me to grow—forget it.

My two friends and I are met at the reception desk by Diane, the activities coordinator. She leads us down the hallway to her office where we are handed a box of finger paints and some paper. "Take these into the activities room and set them out on the tables. I'll make an announcement that you girls are here to do the art project for today. By the way, I have never done finger painting with the residents before, and some of them may not respond well—you know, some of them may have 'issues' with getting the paint all over their hands." With these words of warning, my two friends and I take a deep breath and walk into the activity room.

Here goes nothing!

As I begin to unload the paints from the box, I look up and notice that my two friends are not right by my side. A wave of panic sweeps over me as I see an elderly man come around the corner from the hallway. He slowly shuffles into the room and sits at one of the tables next to me. I look around in hopes of spotting one of my friends—the ones who left me here all alone to fend for myself in this foreign territory.

We will definitely be having a talk on the way home about this running off and leaving me by myself. Do they think I can do this alone?

"What are you doing here today?" the man asks quietly. His speech is slow and sounds as if he has to put forth great effort just to get the words out.

"We'll be finger painting today. Would you like to make a picture?" I figured it wouldn't hurt to just ask. That must be why he came down anyway.

"No. I haven't finger painted in about 50 years," says the man after a bit of struggle. "I didn't like it then, but the teachers made me do it anyway. Now that I am old, I can do whatever I want, and I don't want to finger paint."

Oh no, I offended him and upset him by asking about painting. He probably wants to be left alone, and I am here trying to get him to paint pictures. What is taking my friends so long? What am I supposed to do now? Maybe if I just sit here, someone else will come along who wants to paint.

I look up to see the man facing me. "My name is John. Do you like photographs?"

He removes the button-encrusted baseball hat he is wearing and pushes it across the table toward me. "I take pictures of different things around here and make them into buttons. I sell them at craft shows or to the residents. Do you want to see more?"

I am floored. One minute I think that I have offended this man by asking him to finger paint and the next thing I know, he is offering to show me his button collection. I can feel myself starting to relax. Suddenly it doesn't seem so urgent that my friends come right back.

"I would love to see more buttons," I reply as John is already pushing back his chair. While John walks away from the table, two women come into the room with curious looks. They survey the scene and sit down at the table where I have set up the painting supplies.

"What do I do?" asks the older of the two women. She wears a shirt that says "Mary."

As I reach across the table to demonstrate, the second woman watches me intently. I show them how to stick their fingers in the paint jar and smear it on the page, all the while cautious of how they might react to the new experience. With a slightly shaking hand, Mary sticks her fingers into the bright pink paint and begins to create. Her face looks tense as if she is not sure how to react to the sensation of sticky paint on her previously clean hands. Ever so slowly, a smile begins to creep across her face and her eyes have the slightest hint of a twinkle.

"Look! I'm making a pretty picture," she proudly states to everyone in the room. As I lean down to congratulate her on her budding artistic talent, Mary quickly stands up and begins to clap her hands (which are covered in fuschia paint) with great enthusiasm. Before I realize what is happening, I am caught in an enormous bear hug from Mary and her sticky hands.

"This is fun. You're my friend," she says in a loud voice.

Mary's enthusiasm is contagious. Soon I am up to my elbows in brightly colored paint. I can't help but laugh when Mary gets some of the other residents to get their hands dirty as well by telling them that the paint feels like the Jell-O they had for dessert at lunch that day. Time seems to fly as more people come into the room to participate in the activity and to check out "the girls from Saint Mary's." The next thing I know, it is time for the three of us to leave. We have been here for more than two hours. As we clean up what looks like the ruins of a tornado, Diane comes in to check on us.

"I am very impressed with your work today," she says. "That was the most people we have ever had down here for any activity. The residents must really like you."

Wow! I can't believe I was so worried about coming here!

By the third week, I feel completely comfortable during our visits and could entertain a room full of strangers on my own. At first it is hard to step outside of my own world and into someone else's. After all, people with mental illnesses like schizophrenia and paranoia can't be anything like me or my friends. Or can they?

Mental illness can strike anyone, and the age of onset is usually in the early twenties. Just a bit more or less of one brain chemical keeps me from "going over the edge." It is scary to realize that many of the residents lived relatively normal lives until they developed mental illness. Many of them attended college and held jobs until something happened and they could no longer live without the aid of full- or part-time staff. They seek the same things out of life that I do—acceptance, love, companionship. Seeing the residents' faces when someone shows an interest in their hobbies or just sits back to chat with them helps me to see this. I have seen the same desires in myself, my friends and my classmates.

Each person has something to offer everyone else, no matter how small it may seem. I am constantly reminded of the children's story "Stone Soup." Every person has a gift. Richie can always be counted on to bring a smile to everyone's face when he comes in the room. Doug has a seemingly endless supply of magic tricks that entertain the residents for hours on end. I know I can look to Mike for encouragement—he always has something positive to say.

Then there's Paul. He comes to the activity room religiously at 2 o'clock every Thursday. He saunters into the room and sits at an empty table, all the while in a seeming state of oblivion. Very rarely will Paul acknowledge anyone's presence. He sits at the table, silently rocking back and forth with his eyes closed. When music is played over the stereo, he rocks in time to the beat, and when it is silent, he rocks in time to the beat that must be in his head. Even though Paul never interacts with anyone, he is missed when he does not come to the activity room. What could Paul possibly have to offer anyone? He can't play the piano or draw a picture. He can't even tell a funny joke. I search and search for something that Paul has to offer that would validate my theory that everyone has something to "bring to the table."

One day a resident invites a friend to visit. Debbie walks into the activity room arm in arm with a young woman, goes around the room and introduces each person to her friend. "This is Kay," says Debbie. "She writes articles for the newsletter that are really good. Ernie over here does math all day long on his calculator and can figure out any problem you give him."

One by one she mentions something that each resident can do to make the person stand out from everyone else. Finally, the moment arrives. Everyone in the room has been introduced—except Paul. I start to fidget in my seat. What can she say about Paul? Certainly, he can't do math like Ernie—he probably never even picked up a calculator. Paul's mental illness is so advanced that even I begin to wonder how he can be seen as a contributing member of the community. Without hesitating, Debbie pulls her friend over to where Paul sits in his chair, as usual, oblivious to everyone around him. Suddenly, I know what Debbie will say about Paul, the words I know she will say, the words she has to say.

"This is Paul," Debbie says. "He rocks."

Chat Room

1. Allowing the reader to scan the story is another purpose of answering the 5 Ws + How in a news story lead. What would the reader's likely impression be if this story had a news lead and the reader chose just to scan the text or to infer the story's meaning from just the lead?
2. A *news feature* is tied (or pegged) into a current news event, whereas a *timeless feature* can be printed several days or even months or years later. Which type is "Making Stone Soup"? Give evidence for your answer.
3. Carefully chosen descriptive language may reveal the writer's point of view. Analyze such language choices as *encrusted, foreign territory, chat* and *saunters*.
4. What is this feature's angle? What is its human interest appeal? How does the story evoke emotion? How does the writer's experience lead to universal insights about others' lives?

Create

1. Molly has a distinctive personal style. Imitate her diction and style of phrasing in the first five paragraphs but change the subject to a situation in which you have experienced conflict.
2. Generate a list of ten participatory story ideas.
3. Molly doesn't describe the story's setting after the introduction. Rewrite a couple paragraphs, adding detail you have observed in institutional settings.
4. Watch the film "One Flew Over the Cuckoo's Nest." Look for individuals and situations similar to those in "Making Stone Soup." Write several descriptive paragraphs based on your observations.

The Professional Writing Portfolio

My university's career center called me with the good news. I would be interviewed for a reporter's position with a daily newspaper, and the interview was set for the next day. The bad news: I needed to bring along my writing portfolio.

My what?

No one had ever told me that professional writers maintained portfolios. I had no idea what one should be like. Was it like a model's portfolio but full of writing samples instead of personal portraits? I went out that night and bought a photo album with 8 1/2″ × 11″ sticky pages with glassine overlays. I sat on the floor with files full of college papers and published poetry, fiction and newspaper stories, trying to decide what I should include and to fit material onto the sticky pages in some professional-looking way.

It was a random decision to include my short stories, but I was glad I did. In fact, my interviewers pounced right on them, fascinated. You can't predict what interviewers will look for. I got the job—my ability to write compelling fiction sold me as a journalist (go figure).

Later, I discovered that a writing portfolio was as important for public relations interviews as it was for newspaper.

As the starving Scarlett O'Hara raised her fist and vowed, "I will never be hungry again!" in "Gone With the Wind," I vowed never to be without a writing portfolio again. I now maintain not one but two of them, and I require my students to produce one at the end of the semester. The feedback has been phenomenal. Many have reported how impressed interviewers were that they even had portfolios, hardly expecting the variety of writing samples displayed in them. In a competitive job market, sell your writing skills. Showcase them. Create a knock-their-socks-off portfolio. Here are some how-to tips:

- Buy glassine pockets that fit into a three-ring binder; place your writing samples in the pockets to keep them from getting dog eared and discolored.
- In the three-ring binder, organize your samples in a logical way. For example, you might divide sections by courses or writing types, have a section for work done as an intern, or have a section for works published in the college newspaper.
- Place your résumé first in the binder (include extra copies of it in the pocket—handy in case you walk into an unexpected multiple-interview situation).
- Make a Contents page to list the categories of writing inside.
- Purchase dividers or create your own to guide people to the section that most interests them.
- All writing samples should be on white paper, but you may get creative with the section dividers—use color and graphic designs to reflect your personality or the image of yourself you wish to communicate. Include blurbs that detail the assignment (prospective employers are interested to know what you were asked to do and how you did it).
- Have someone with expertise in your chosen field evaluate your portfolio and include this evaluation in the portfolio; include internship and other employment evaluations.
- Place multiple revisions underneath the final versions of your writing samples; prospective employers will be interested in your ability to edit and revise.
- The quality of your writing matters a great deal. Don't display course papers with

numerous professorial edits. Revise everything so that it is perfect—clear, concise and correct.

- Show your flair, but keep it professional. Above all, be neat and organized in presentation.

The second portfolio I now maintain measures 16″ × 19″, the perfect size to display newspaper pages. In my case, I wrote a number of front-page and magazine features that I wanted to show as they were published, unfolded and uncut, with the accompanying headlines and art (photographs and graphics). If you have worked on page layout and design, or have created the graphics or taken the photos, you should also consider displaying the pages intact.

Some students have felt so strongly about the portfolio as a worthwhile investment that they have purchased relatively expensive, zippered portfolios made of leather as a showcase for their work. Others have created duplicate portfolios and mailed one ahead prior to the interview. A recent trend is to present writing samples on a computer disk or CD-Rom, a format less expensive to create, duplicate and mail ahead.

However you decide to showcase your work, get busy now—don't think about it tomorrow, even if Scarlett O'Hara argues that tomorrow is another day. Remember, her procrastination caused her to lose Rhett Butler. As your professor returns features to you, revise them and get them right into those glassine pockets. Former students tell me that a great writing portfolio will not only help land a first job but also give a competitive edge for a promotion or that move to a better position elsewhere. Update your portfolio throughout your college career and your professional life.

FEATURE STORY

"The Internship Payoff"

Nellie Williams started as a student intern, then was hired by the newspaper as a staff writer/photographer. Curious about internships—and what it takes to be hired as an intern—Nellie interviewed Margaret Fosmoe, internship coordinator for a regional newspaper.

Newspaper recruitment officers receive hundreds of applications for positions each year. So what can you do to stand out from others in a pile of résumés?

Margaret Fosmoe, internship coordinator for a regional newspaper, knows firsthand that internships are key to landing that first job. As an undergraduate, she interned during two academic semesters and two summers before she was rewarded with her first newspaper job in Kalamazoo, Mich.

"I kept in touch with people and networked," said Fosmoe.

But internships are not only important for connections. They are also essential in developing important skills that are impossible to learn in a classroom and in gaining a variety of experience across the industry.

Students don't necessarily need to be journalism majors to apply for an internship. However, they need to have some background in reporting.

"It is an extremely competitive market," Fosmoe says. "Students need to demonstrate an interest in journalism. Editors want clips or some sort of evidence that a student has sought out experience before."

Fosmoe advises students to "try student publications, seek out freelance work and apply for an internship. If you can't swing an internship, try freelancing. Get a few clips to add to your portfolio."

Fosmoe knows that the job market in today's economy is tough, and many publishers are doing away with summer internships. Still, she advises students to write and to create a writing portfolio.

Newspapers typically offer internships for reporters, photographers and page designers. "This year we had a record of 130 applicants—it was absolutely overwhelming," Fosemoe said.

So what are editors looking for when they decide who will receive the internship?

"Editors want students to be multifaceted. A fair amount of applications we receive are so specialized that students almost market themselves out of the job market," Fosmoe explained. "Editors are much more interested in students who have reporting skills."

Fosmoe urges students to read newspapers and magazines and look for role models, "to find a mentor and to talk with them regularly and ask for feedback."

Most important, don't give up. Keep writing.

⌕ View _____

The purpose of a story may be to inform, educate, inspire, record historical events, entertain and/or transmit cultural mores. All stories, whether news or feature, must generate human interest; typically, they include common required elements: who, what, where, when, why and how. On one hand, hard news stories focus primarily on the reporting of crucial information, with the 5 Ws + How answered in the lead, or first, paragraph. Hard news stories use inverted pyramid structure to support the lead paragraph, in descending level of importance, without adding personal reflection or opinion. On the other hand, the writer's point of view determines the feature story's angle. Feature writers maintain a story's information level by evenly distributing the 5 Ws + How throughout the story, yet they have more room for reflection, emotional appeals and personal style.

The student writer should create and maintain a portfolio of writing samples. The portfolio may include revised class assignments and published works. Above all, the student writer should read features available in a variety of sources—newspapers, magazines, online sites that publish prize winners and, of course, this book.

📖 *Help*

angle   the writer's point of view—the person's "take" on, or approach to, an assignment, as reflected in the story idea.

inverted pyramid style   the custom of beginning news stories with a lead that answers the 5 Ws (who, what, where, when, why) + How, then offering information in descending order of importance. This ordering creates a story structure that resembles an upside-down pyramid.

lead   the first paragraph or paragraphs of a feature story that must attract readers' attention, pull them into the story and create an atmosphere or visual picture. These paragraphs *lead* readers into the story.

news feature   an accountant that casts news into the more creative, reflective structure of the feature story.

participatory feature   the story that results when writers participate in an activity or event that takes them outside their own experience to learn about the experiences of others.

story   an account, written or verbal, possessing a beginning, middle and end, that essentially derives from common purposes that have not varied much throughout human history: to inform, educate, inspire, record historical events, entertain and transmit cultural mores.

Notes

1. Bob Steele and Roy Peter Clark, "Coverage of the Kennedy Crash: The Crisis of Celebrity Journalism," The Poynter Institute, 3 Jan.1999<http://www.poynter.org/special/point/Kennedy.htm>.
2. Meghan Martin, "Journalism Panel Relates 9/11 Changes," *The Observer*, 17 Sept. 2002, Vol. XXXVII, No. 16, 4.
3. W. Richard Whitaker, Janet E. Ramsey, and Ronald D. Smith, *Media Writing* (New York: Longman, 2000) 153.
4. Chip Scanlan, "Writers at Work: The Process Approach to Newswriting," Poynter Online, 4 Jan. 2000 http://poynter.org/special/tipsheets2/reporting.htm.
5. Rives Collins and Pamela J. Cooper, *The Power of Story* (Boston: Allyn & Bacon, 1997) 1.

2

Feature Writing as a Process

All writing is ultimately a question of solving a problem.
—*William Zinsser in* On Writing Well[1]

Troy Donahue was coming to town. Sure, I knew exactly who he was. A member of the baby boom generation, I was there in 1959 when Donahue first appeared on the silver screen, tall and handsome and blond, in "A Summer Place." He wore white pants and a red sweater and came to represent everything desirable *and* acceptable in an American male of the era. In the film, he made just one unforgivable mistake: He gave in to lust in a sexually repressive society.

Decades later, my editor asked me to do a *lead story* about Donahue for the entertainment section of The South Bend Tribune. He was to appear onstage in South Bend in the touring production of "Bye Bye Birdie." The actor's press agent set up a telephone interview, and the newspaper received color slides of scenes from the play. The editor assigned a *personality profile*. What might have seemed an easy assignment actually presented a major problem. I knew who Troy Donahue was, but would my readers? Unless they had caught "A Summer Place" on the late-night movies, how could the audience relate to him?

I had a problem.

As Zinsser suggests, all writers begin with a problem. Solutions occur as writers work through the writing process. Considerable time should be spent in idea generation, research and organization before the later stages of writing and revision. While news writers work quickly on deadline, the feature writer has greater lead time—and ought to use it. A good feature will not be crafted overnight. As Brendan Hennessy has said, "Often, the easier a feature article is to read, the harder it has been to write."[2] The key word is *read*, as in *reader*. Throughout the writing process, feature writers must keep the reader foremost in mind.

As I began work on the Donahue story, the first step of the process faced me: generating the idea. I had the assignment, but an assignment is not the same as an idea. I wanted an angle that would intrigue a diverse newspaper audience with a late '50s matinee idol. I could count on the fact that the older generations, who patronized '50s movie theaters or watched late-night movies, would recognize "A Summer Place" and its rising star. However, my readers would also include younger people who might not.

To resolve this problem, I thought of another Donahue-related vehicle that young people *would* recognize—the 1970 musical "Grease." The musical connects Donahue to its theme of '50s psychomachia (the playing out of the decade-long battle between the "good" and "evil" elements in American society) through the lyrics of one of its songs, "Look at Me, I'm Sandra Dee," which ridicules the good girl/good boy ideal. "Grease" solidified Donahue's status as an icon; he came to symbolize the timeless struggle between teenagers and adult morality.

Once the writer has the idea, the research stage of the writing process begins. Prior to the telephone interview with Donahue, I rented the video of "A Summer Place," a film I had watched before, though not from the perspective of a reporter observing a person to be profiled. Novice writers often rely on distant memories to conjure up story details; that's a big mistake. Always take a fresh look. Other research sources included publicity materials provided by the touring company and Web sites devoted to Donahue and his movies. I watched film versions of "Bye Bye Birdie" and "Grease," then drafted questions for the phone interview. In the interview, Donahue talked about his other 1959 film, "Imitation of Life." I realized I'd have to re-view "Imitation of Life" before going on to the next stage of the process.

As I outlined the story and gave it *context*, my personality profile took on aspects of other profile types. The emphasis on the quirk of an era, the puritanical mindset of the 1950s, expanded the feature's focus from Donahue the person to the turbulent time in the United States that he came to represent. A feature writer may *profile* a person, place or thing, and the story I was writing began to encompass it all. Don't be afraid to let the story idea refine in the writing process; the idea just begins your work.

Idea Generation

Story ideas exist everywhere. Writers generally have a "nose for news," a natural curiosity about people, places and events. The 5 Ws + How describe the skeleton for a story idea—who did (or does) what, where, when, and why? As you've noticed, in regard to the Donahue story, an assignment won't fully express the story idea. The word *assignment* derives from the Latin for "to mark" and denotes something more tangible than *idea*, which in Latin means "a notion." In a sense, the editor marks the writer down for a story (the assignment) about a person, place or thing. The type of feature might also be specified. The angle, the story idea, reflects the writer's notion, or point of view, about the assignment.

In some cases, the story may not derive from an assignment; a writer may pitch a story idea to an editor. In either case, the writer initially generates the idea.

The idea for a story I wrote about a Miss America finalist first occurred when I saw her perform in a dance concert I was reviewing. She struck me as a person of extraordinary energy and talent. Months later I read in the newspaper that she had been named Miss Indiana. News stories must be timely, and features also tend to be pegged to timely events. I pitched a story about her to my editor, who scheduled the story to appear in a timely fashion—a month before she would appear in the Miss America competition.

Almanacs and fact books may generate story ideas. For example, the Advertising Age Ad Almanac would be a good source for trend feature topics. According to the Ad Age Almanac, the United States is 33rd in wine consumption in the world,[3] whereas beer is a popular American drink with Bud Light the number one beer in the U.S. market.[4] A feature on the American tendency to drink beer rather than wine would address the obvious question: Why?

Feature writers sometimes peg stories to anniversaries. For example, on March 11, 2002, the six-month anniversary of the 9–11 tragedy, the media blitzed the nation with stories about the attacks on the World Trade Center. Some critics claimed the media rushed to create the event, not waiting the usual year to observe the anniversary.

Some ideas may counteract news events. For example, the release of a report on the health risks of obesity might prompt an occasion to write a feature about someone who has battled anorexia.

Seasonal stories abound. Newspaper sections and magazines targeted to homemakers typically run prominent stories on foods and decorations related to specific holidays.

Not all features peg to news events, of course. While the Donahue story was prompted by the celebrity's appearance in the community, Troy Donahue carries enough news value to be profiled even without a current news peg. And personality profiles don't have to focus on celebrities. The night cleaning crew at your college might include interesting people who have stories to tell even though few students have had the opportunity to meet and know them.

If you write for a local or regional newspaper, your idea must tie to the community. The *local tie* that establishes *proximity* must be made clear by the second or third paragraph of the finished story. As a writer for a regional newspaper, I quickly came to expect two questions from the editor whenever I pitched a story: *What's the angle*, and *What's the local tie*? Had Donahue not been scheduled to appear in the newspaper's circulation area, the editor might have refused the story idea. On occasions when I've had the opportunity to interview a celebrity and could not find a local tie, I looked for other publications to market the story. Writers communicate story ideas in query letters, a subject discussed in Chapter 7.

Your personal experience may turn up numerous story ideas. Chapter 1 features stories generated by unusual personal experiences. The business profile in Chapter 8 resulted from my curiosity to learn more about my hometown "sweet shop." When I visited my hometown years ago, I noticed that the old teen hangout, Marz Sweet Shop, was still in business, although it had a different clientele. I'd call this a nostalgic idea.

In "Professional Feature Writing," Bruce Garrison lists additional sources for feature ideas: personal and professional contacts, college/university campuses, meetings and conventions, readers, editors, other writers, community centers, bulletin boards, calendars and datebooks.[5]

This list reminds me that my first published magazine feature resulted from my interest in a writers' conference to be held at Indiana University, Bloomington, in the summer of 1975. The local IU campus calendar listed the conference, and I noticed that some big-name writers would teach workshops. I knew why *I* was interested; I was a young, aspiring writer. I had to ask why the readers of a regional magazine would care about the conference. I had to shape the idea, choose the right angle, for my audience. It occurred to me that writers carry a certain mystique: What do they do? How and why do they do it? I decided to take nonwriters to the heart of what to me has always felt like a performing art with an absentee audience. Oh, the readers are out there, all right. You might hear from them via a letter to the editor about one of your stories, or you might run into them at the grocery store. Readers make up a tangible audience, and even though you can't see them while you're writing, they're the reason you're doing this, from idea generation throughout the entire process. Explaining that notion to readers became the story idea.

Natural-born storytellers work 24 hours a day. There's no such thing as going home from the job. They always observe and listen, even in their sleep. The biblical Joseph made a living telling stories he literally dreamed up. Joseph tailored his stories for his audience—Egypt's Pharaoh. Joseph had to ask whether the story would be of interest to his audience, and how the story might be relevant—what aspect(s) would best connect with audience interests at a particular time or at any time. If you've seen the musical "Joseph and the Amazing Technicolor Dream Coat," you know that Joseph's stories touched on the Pharaoh's concerns about Egypt's impending famine. Idea generation finally requires the storyteller to decide whether and how this idea might engage the audience.

Research

Once the writer has an assignment and an idea, the research phase of the process begins. The attempt to write the story before the research has been done constitutes a pitfall for novice writers. Research backbones all media stories, whether news or feature, newspaper or magazine. Resources for writers include subscription information sources (for example, wire services) and free information sources (for example, Web sites and press releases), library books and periodicals, observation and interviews. The research phase occupies its own section of this textbook, testimony to its importance. The next chapter summarizes research options for feature writers, and the following chapter covers the art and science of interviewing. You will learn more about research as you read on in this book.

Organization

When the writer has gathered a body of information, facts and quotes and has observed detail through research, the process of organization begins. The writer faces the problem of how to best organize essential material and what to discard. A young writer may be so attached to research materials that it is painful to discard the inessential. After all, this is a cool fact, and I worked hard to unearth it. Sound familiar? I know the feeling, but you have to get over it.

In "Problem-Solving Strategies for Writing in College and Community," Linda Flower suggests that writers create "issue trees" to sketch out ideas in a logical order.[6] She suggests finding a key word or phrase for each idea generated for the story and placing the key words hierarchically in a hand-drawn tree. The supporting ideas then visually branch from the central idea for the story, helping the writer organize and reorganize supporting facts and concepts. Giving visual form to fleeting thoughts seems to be the essence of story planning.

The technique I've come to favor over decades of media writing is *clustering*. I gather my story materials and select the details that will illustrate my central idea (angle). I circle paragraphs in press releases and other background materials; I mark quotes and observations I believe are important in the notes I have taken. I often identify a lead in a quote or observed detail that I have circled. Then I analyze what I have marked and begin to cluster. Although I didn't save my notes, the clustering that preceded my writing of the Donahue story would resemble Figure 2-1.

Some writers create clusters of ideas before writing an outline. Other writers bypass clustering and go straight to outlines to resolve the organization problem. An *outline* is a plan or scheme for a story's main points and details. You have probably learned to write formal and informal outlines in your English classes. A *formal outline* requires

Idea: Troy Donahue became an American icon, although not the type of icon he expected to be.

Donahue admired Brando & Dean
He didn't become a rebel like them
He became a "goodie two shoes"

"Goodie two shoes" son in "A Summer Place"
"   "   "   father in "Birdie"
Wears same red sweater in each

Icon of the '50s
Represents "good"
"Birdie" spoofs '50s
"Grease" spoofs '50s
Rizzo suspects sexual repression
    behind Donahue's "goodness"

"A Summer Place" a cult classic
Donahue's TV & film credits
"Imitation of Life" works against typecast
Observed detail: Donahue backstage

FIGURE 2-1 *Donahue Story (Clustering Ideas)*

the writer to put down ideas in full sentences and paragraphs. An *informal outline* does not require that details be expressed in complete sentences and thoughts. An informal outline for the "Troy Donahue" story, which follows in this chapter, appears in this chapter's Click Here.

While many writers work from formal and informal outlines, and many teachers require that students write and submit them, not all writers work the same way. Some writers "mull" their ideas over a period of time before putting pen to paper (or fingers to computer keys). Arthur Miller has said that his great masterpiece of American theater, "Death of a Salesman," came to him as images generated by his subconscious mind. Many of these images derived from his observations of people and their lives: "a little frame house on a street of little frame houses, which had once been loud with the noise of growing boys, and then was empty and silent"; "the cavernous Sunday afternoon polishing the car"; "so many of your friends already gone"; a private man "in a world of strangers."[7] He watched the action of the play unfold in his mind long before he was able to make sense of the images and get the play down on paper. Like some other great writers, he felt the play wrote itself. Of course, we are not all great writers, and even if we were, not every story we write will be a masterpiece.

So we outline.

Carrying an idea in your head until it becomes a coherent plan is a luxury few media writers could indulge anyway. Arthur Miller wasn't working on a newspaper's deadline. Whether the plan for the story takes the form of a sketched tree, a clustering of ideas, a formal outline, an informal outline or simply an organization of research materials by writing numbers or letters on the materials themselves, *clearly the writer must research and plan the story before starting to write it.* This is a known fact, a universal truth shared by practical writers, not some academic fabrication.

Context

Feature writers do not work in a vacuum; they write for readers. This means the feature writer must be sure that a story conveys meaning to its readers. *Context*, from the Latin *contextus*, means "to join together." A feature story's social context gives meaning to the event or situation depicted; it joins the story's details to a larger sphere of meaning.

For example, Molly Donnellon, in her feature in Chapter 1, "Making Stone Soup," establishes context when she tells us that her story takes place at Portage Manor, and that Portage Manor is "a residence home for adults who are mentally ill or handicapped." We know where we are. She tells us that she is volunteering there, that this is her first day and that she knows nothing about people who are mentally ill, so we understand her apprehension. She introduces a variety of people but carefully gives context to each person; for example, she tells us that Diane is the activities coordinator, John likes photography, Ernie is a math wiz. Molly's giving context to the reader finally creates an exciting ending for the story. Paul seems to defy context. His existence seems meaningless until Molly gives words to his function: he *rocks*, which we understand as a play on words, capitalizing on a well-known, contemporary expression. Molly's context connects the detail about Paul to a sphere of larger social meaning.

The reader's need for context must be addressed. The writer should let the reader know where and when the story takes place, briefly introduce each person at first mention, provide sufficient background so the reader can understand situations and events, and let the reader know who you are and why you're there in the event that you, as the writer, enter into the story.

Drafting and Revision

If the outline plans the story, the *draft* forms the basis, or blueprint, from which the finished story will emerge. *Revision* refers to a series of refinements as the writer works to perfect the story's structure and language. However, professional writers do more than just *local revision* (proofreading, correcting errors, polishing sentences and paragraphs). According to Linda Flower, experienced writers revise globally "with the goal of expanding, reorganizing, or changing not just words and sentences, but the gist of a paragraph, an argument, or even a whole paper."[8] *Global revision* requires a clear purpose in writing the story and a gist of what the writer wanted to say.

Get the draft written fast, then let it sit at least overnight (that means writing ahead of deadline, not pulling an all-nighter the day before the feature is due). Now approach the story as a reader might and ask the questions a reader might ask. Revision then involves reworking the story to answer all the reader's anticipated questions. Student writers should seek reader response throughout the revision process. Ask your roommate to read your feature; ask your girlfriend or boyfriend to read it; ask your mom or dad, sister or brother. Have one reader proofread; have another tell you whether the story was easy or difficult to read (does it flow?). Work collaboratively with others in your class. Ask your teacher to go over a draft with you.

One of the problems novice feature writers frequently encounter is how to create emotion and immediacy, to be "present" in the story, while maintaining journalistic objectivity. Nellie Williams faced this dilemma with her first feature story. She had the opportunity to interview a mother who had given her daughter up for adoption yet was allowed visitations. The daughter was present for the interview, and Nellie felt compelled to refer to herself, as the writer, as she described the dynamics that occurred when the three gathered at a Minneapolis coffeehouse.

Students often ask how they can shift focus from themselves to their subjects and eliminate the first person, the *I*, from their features. Look at how Nellie revised her lead to achieve this purpose.

Original Lead
It is a cold, crisp Saturday afternoon. The steam from my spicy chai tea cools off my rosy cheeks. Leslie Cook* sits across from me putting a puzzle together with her four-year-old daughter Rachael. Rachael is laughing. Some of her blond curls have escaped out of her pink barrettes and frame her face. She has bright blue eyes just like her mom. To any passerby, one would see the resemblance and think that it was a normal mother and daughter enjoying a lazy afternoon at a local coffeehouse in Minneapolis. However, Rachael and Leslie have anything but a normal relationship.

*"Leslie Cook" is a pseudonym.

Nellie needed more specific details: What afternoon? What coffeehouse? Her lead was wordy, and she footnoted the mother's pseudonym (feature stories rarely employ footnotes). She had word choice problems, especially in referring to the mother and daughter as not "normal." And she needed to take herself out of the story. Two versions later, Nellie's lead read like this:

Revised Lead

It is a cold, crisp Saturday afternoon in January. Leslie Cook, who asked that her name be changed for privacy purposes, is putting a puzzle together with her 4-year-old daughter Rachael in The Spy coffeehouse in Minneapolis. Rachael is laughing, and some of her blond curls have escaped from her pink barrettes, framing her small face. She has bright blue eyes just like her mom. Any passerby would see the resemblance and assume it was an ordinary mother and daughter enjoying a lazy afternoon together. However, Rachael and Leslie have anything but an ordinary relationship.

Nellie also brought herself into the body of the story. Notice her problems in the original paragraphs, followed by her successful revisions.

Original Body Paragraphs

Leslie informs me on the different types of adoption.

I watch as Leslie gives Rachael another piece of the cookie. Leslie is young, about 21. I could not imagine loving someone so much that I would give them up to have a better life. It is a very unselfish thing to do.

I hold Molly in my lap before Rachael grabs her back. Leslie asks Rachael how her doll has been feeling. "She was sick last week, but mommy gave her medicine," Rachael explains to Leslie, her birth mother. Leslie does not seem to mind that she calls her adoptive mother "mommy." I ask Leslie how it makes her feel.

"She is lucky," Leslie explains to me. "She has two 'mommy's'; one that can nurture her daily, and one that will love her from a distance."

I am amazed at how at ease Leslie is with the adoption.

Revised Body Paragraphs

Leslie explains the different types of adoption.

She gives Rachael another piece of cookie. Leslie is young, about 21. Although she made the sacrificial decision to give her daughter up in adoption, she chose the open type in which information is fully disclosed and meetings between birth parent and child can be arranged.

Rachael slips out of Leslie's lap and grabs the doll that she had discarded earlier from the floor. Leslie asks Rachael how her doll has been feeling. "She was sick last week, but Mommy gave her medicine," Rachael explains to her birth mother. Leslie does not seem to mind that she calls her adoptive mother "Mommy."

"She is lucky," Leslie explains. "She has two 'mommies,' one that can nurture her daily, and one that will love her from a distance."

Leslie seems amazingly at ease with the adoption.

As Nellie learned, the feature writer recaptures the scene but should seldom star in it. Nellie did more than change the words; the global revision shifted focus to better achieve her purpose—to tell the story of how a birth mother and her daughter adjust to and benefit from an open adoption. The "gist" of what Nellie wanted to say became more directly conveyed to the reader with each of her three revisions.

How many drafts will it take to reach the final, finished story? That's hard to predict. My answer would be, "As many as it takes." In an interview with George Plimpton, Ernest Hemingway talked about rewriting the last paragraph of "A Farewell to Arms" 39 times. What was the problem? Getting the words right. Hemingway revised until he had solved his writing problem; he had professional tenacity. So can you.

Click Here: Outline for the Donahue Story

Idea
Troy Donahue has become an American icon, but not the icon he set out to become.

Local Tie
Appearing in "Bye Bye Birdie" in South Bend on March 18

Supporting Detail
A. Donahue as American icon
   1. A "good" boy in "A Summer Place"
   2. A "goodie two-shoes" father in "Birdie"
      —his songs in "Birdie" spoof "A Summer Place"
      —his costumes for "Birdie" like "A Summer Place"
B. "A Summer Place" establishes him as icon
   1. Parallel plot—teen romance contrasts parents' adultery
   2. The musical "Grease" spoofs "A Summer Place"
      —brings Donahue to a new generation
      —satirizes the '50s
      —Rizzo's song
   3. "A Summer Place" has cult following
      —June 1997 special showing in New York City
C. Donahue has other film and television credits/typecast by "A Summer Place"
   1. Lead roles in television and film
   2. Antihero in "Imitation of Life"

Ending
Donahue backstage: handsome, polite, "good"

Troy Donahue

Following the initial telephone interview with Troy Donahue, I attended the opening performance of the South Bend run of "Bye Bye Birdie." Afterward, I went backstage to meet the man I had interviewed. A telephone interview gives the writer no opportunity to observe the person, and I wanted to flesh out the feature with personally observed detail. I'm glad I did. Although Donahue was much older, he was still the somewhat hesitant and handsome boy who had charmed so many of us in "A Summer Place." Stage acting just didn't do justice to the long, shy stare that made his face so appealing in the movies. In retrospect, I could have compared Donahue's gaze to the disarming gaze of England's Prince William. That just didn't fit the focus of the story, though. The feature ran on the front page of The South Bend Tribune's WKD (weekend) section on March 19, 1998.

Like most teenage boys in the 1950s, Troy Donahue admired the artistry of two rebels with talent—film actors Marlon Brando and James Dean.

Unlike most teenage boys, Donahue ventured to Hollywood to pursue the paths of his heroes. He had just graduated from New York Military Academy.

Instead of becoming the next Brando or Dean, Donahue became another kind of American icon, which he described in a recent telephone interview as the "innocuous, all-American, goodie-two-shoes guy."

In the 1959 film "A Summer Place," he played Johnny, the heartbreakingly handsome, polite, middle-class boy who falls into the unthinkable predicament of fathering a child out of wedlock with Molly, played by Sandra Dee. He made his screen debut in a preppie sweater and white pants. He couldn't have distanced himself more from Brando and Dean, whose ripped T-shirts and disrespectfully tight jeans had shaped their rebel images.

While "A Summer Place" took the problems of Puritanical repression quite seriously, it wasn't outwardly rebellious. Just a year after its release a new musical lampooned the same prejudices and Ed Sullivan–variety family values. "Bye Bye Birdie" opened on April 14, 1960; its 608 performances became one of the longest Broadway runs of the decade.

Ironically, Donahue, the goodie-two-shoes son in the 1959 film, plays a goodie-two-shoes father in the national touring company production of "Bye Bye Birdie." At the time of the telephone interview, Donahue and the musical were en route to South Bend for three shows.

The musical's creators transparently modeled its central figure, Conrad Birdie, after Elvis Presley. Inducted into the army, just as Elvis was, Birdie needs a public relations strategy to keep his stardom aloft. The "save" comes from his songwriter-manager Albert Peterson, who writes a ballad about a farewell kiss Birdie will plant on one of his fans. This will all take place on that sacred cow, "The Ed Sullivan Show" (remember that, in reality, Sullivan's cameras shot Elvis only from the waist up to maintain '50s decorum and decency).

In the touring production, Donahue plays Mr. MacAfee of Sweet Apple, Ohio, the frenzied father who idolizes Ed Sullivan and whose daughter Kim has been selected for Birdie's farewell kiss.

When Donahue laments the current decline of contemporary youth in his "Kids" number, he's on the opposite end of the same dilemma posed in "A Summer Place." His other song, "Hymn for a Sunday Evening," spoofs America's allegiance to Sullivan's proper presentation of televised entertainment.

Costumes designed by Miles White for the Broadway production of "Birdie," especially Birdie's garish, skin-tight gold lamé pants and jacket, were meant to shock the puritanical establishment that carried over into the new decade of the 1960s. The costumes for the touring production are "very authentic," according to Donahue, "all chartreuses and other '50s colors," men with sideburns and slicked-back hair, women wearing bobby socks and snug sweater sets.

Donahue wears his signature red sweater, made famous in "A Summer Place." The real-life father of two and grandfather of three, Donahue calls MacAfee "a fun role. Kids and parents have always been at odds, and this play makes fun of the generation gap."

But the generation gap as depicted in "A Summer Place" horrified teenagers in the 1950s audience who identified with Johnny and Molly, and likewise grappled with the repressive, post-McCarthy, pre–civil rights mindset of the nation. Surprisingly, Donahue always thought the movie was "very funny. The clichés were so incredible."

The parallel plot of "A Summer Place" contrasts the teen romance of Johnny and Molly with the rekindling of an old affair by Johnny's mother (Dorothy McGuire) and Molly's father (Richard Egan), both miserable in bad marriages. Turned away by Molly's uptight New England mother and Johnny's alcoholic father, the two run away to get married. They are united by a seedy justice of the peace, who eyes the couple suspiciously, and spend their wedding night in a junker car parked under a bridge. Although appalled by the parental affair, Johnny and Molly have nowhere else to go but to the imagined den of sin where his mother and her father cohabitate. They are welcomed there with open arms and a joyous Hollywood ending, complete with the prize-winning Percy Faith theme song.

Although the satire lacked subtlety, "it went right to the heart of the problem," American Puritanism, Donahue says.

The film made icons of Troy Donahue and Sandra Dee. In 1972 Jim Jacobs and Warren Casey created the enduringly popular "Grease." Ridiculing '50s repression, Sandy (the good girl) goes bad to get her guy, and in the world of this musical, that's good. If naming the good girl Sandy after Sandra Dee weren't enough, one song offers an unabashedly cynical view of the goodness of both Dee and Donahue.

The musical's rebellious Rizzo, the closest you can get to a female Brando or Dean, masquerades in a blonde wig to impersonate and slam her nemesis, the virginal Sandra Dee. In an aside in her song, "Look at Me, I'm Sandra Dee," she directly addresses Troy Donahue. She isn't fooled by his goodie-two-shoes facade; *she* knows what he wants to do. Using Donahue as an icon, the flip side of his own rebel-with-a-cause movie heroes, Rizzo undercuts the '50s double standard—do what you want in private, but cover it up in public. The musical "Grease" appropriated the iconography of "A Summer Place" and passed Sandra Dee and Troy Donahue on to new generations of teens.

In June 1997 Donahue and Dee attended a special showing of "A Summer Place" in New York City. "They packed the place," Donahue recalls. "Some people knew all the lines and said them out loud, along with us on the screen, almost as they do for 'Rocky Horror Show.' "

Donahue has amassed three decades of film and television credits (leading roles on television's "Surfside Six" and "Hawaiian Eye," featured roles in such films as "Parrish," "Susan Slade" and "Godfather: Part Two," and major roles in made-for-TV movies such as "The Lonely Profession" and "Malibu"), but "A Summer Place" remains his sentimental favorite because "it did so much for me."

Still, the film typecast him.

In 1959 he made a brief appearance in the film "Imitation of Life," which Donahue says "rocked the industry and society with four racist words."

Reminiscent of the more recent tearjerker "Beaches," the 1959 film features two struggling single mothers, one white (Lana Turner) and one black (Juanita Moore), who move in together to raise their daughters. The white woman finds fame and fortune as an actress. The black woman remains as her maid; her daughter, Sarah Jane (Susan Kohner) passes for white and falls in love with a white boy, Frankie, played by Donahue.

In his one scene, Donahue as Frankie confronts Sarah Jane with the gossip he has heard and, in anger, repeatedly hits her. "I was afraid to do that part," he says. "I thought I would lose all my black fans. Instead, I gained a lot of black fans for hitting a girl who tried to pass for white."

Despite the social impact of his earlier works, Donahue, 63, says that touring "Birdie" is a "lifetime experience" for him. "I've done stage before," he says, "but never musical theater." He studied with a voice coach, worked with the musical director and got some help from his girlfriend, Zheng Cao, a mezzo soprano.

On opening night in South Bend, Donahue appeared backstage in his signature red sweater, still heartbreakingly handsome and basketball-star slim and tall. James Dean perished in a controversial car accident decades ago. Another of the actors who inspired Donahue, the obese and reclusive Brando, scorns fans and refuses interviews. Always on the flip side of his intended image, Donahue politely signs autographs. When a nervous, aging fan dropped her program, he bent way down to retrieve it, the all-American boy picking up the pieces of a bygone era, the last gleam of chivalry. Or, some might argue, the darkness of moral repression.

Chat Room

1. The writing problem for this story was to reach a younger as well as an older audience. Does any aspect of this story resonate with you? If so, explain.
2. If this story had been given more space, what information would you like to see added?
3. Identify sentences that give the story context.
4. In a group, discuss your usual writing methods. Do you cluster material before you outline? Do you prefer formal or informal outlines? Have you ever "mulled over" a story?

Create

1. Rewrite the "local tie" paragraph. Peg the story to a different publication and time, for example, a national movie magazine in 1960.
2. Generate ideas for stories about entertainers of your generation. Choose the one you'd most like to write. Here's your problem: How will you angle this story so it will appeal to multiple generations of readers? Find an angle that solves the problem.
3. Outline and write a story using the angle you came up with in exercise 2.
4. After you've completed exercises 2 and 3, write a reflection paper about your experience with the writing process.

"Men's Pickup Lines, Then and Now"

In response to the first Create exercise following the Troy Donahue feature, students Renée Donovan and Sara Pendley wrote the following lead. Beginning with that lead, they went on to research and write a trend feature about pickup lines. They observed at a local night spot, interviewed young people and consulted Web sites. The feature updates the Donahue story for a younger generation; it also shows the correct way to cite a Web site in a feature story, according to The Associated Press Stylebook.

"I don't want to make you do anything you don't want to do," rambles Troy Donahue in the 1959 movie "A Summer Place." Donahue's classic pickup line is still used today. Men of all ages, shapes, and sizes use the same old lines to try and woo a woman. The problem with this: women fall for them. The girls of the 21st century are not much different from those of the 1950s. They still fall for handsome looks, savvy apparel and smooth lines.

What makes a guy handsome? Some women find the clean-shaven and well-manicured men attractive, while others find the rugged, outdoorsy type fit their style. However a man looks, he takes time to accentuate his best features to attract that special woman. Sam Gallo, 24, a single man out in the dating scene, provided his insight on the preparation for a night of wooing. "Before I go out, I would make sure I was clean, shaven and smelling as sexy as I could be," he said. "The perfect outfit accentuating my muscles and butt are important in catching the eyes of the ladies, especially if I want a girl to come home with me." Apparently men take time to prepare for an evening "on the town" just as women do. Both sexes have seduction on their minds; however, men desire to take a woman home for a particular reason—sex.

This relates back to the caveman. In theory, the caveman wants to conquer his mate. In many cases, men have not grown from this primitive mindset and are still determined to ensure a sexual encounter. That means some men must compensate for so-so looks or a nervous personality through their apparel. Guys try to look savvy according to their personal style and taste. Whether in flannel or silk, Abercrombie or Bacharach, men dress to attract the next victim.

Savvy dress may draw the female, but to keep the interest or "break the ice," a smooth pickup line is necessary. It is well known to the dating population that the pickup line will either make or break the initial connection. Over the last 50 years, pickup lines have been manipulated to suit individual styles. Men look for the perfect line that will keep a woman's interest every time. The key question is whether the pickup lines have truly changed over the decades. "Pickup lines," http://pick-up-lines-pick-up-lines.com, is a Web site dedicated to the subject. According to this site, the medieval version of a "Hi! I'd like to get to know you" would be "You must be a skilled archer, my lady, as you have sent your arrow straight into my heart." This Web site also provides a twist of humor that a man might use to gain the attention of a woman: "Your body's name must be Visa, because it's everywhere I want to be." Or how about a little romance? The man asks: "Did it hurt?" The woman replies, "Did what hurt?" Answer: "When you fell from heaven."

Karen Shaff, a 21-year-old college student, describes herself as "single and looking." Typical lines such as "Wanna dance?" and "You're hot. Want a drink?" don't seduce Schaff. "When guys show interest by asking me out on a real date or actually ask questions to get to know me, I get interested," she explains.

Men attempt to seduce women at any singles place. Listen carefully. Women will even find verbal seductions in songs they hear every day. For example, in Tom Wopat's 2002 CD release "In the Still of the Night," he sings a version of "Baby It's Cold Outside" with Antoinia Bennett. In this seduction, Wopat makes reference throughout to the beauty of the woman he wants to stay with him for the night. Phrases in Wopat's version of the Frank Loesser song reflect what many women hear from men while in the singles scene. For example, comments about "delicious" lips appeal to the desire many women have to be seen as beautiful in the eyes of the men they adore. When Wopat notes the thrill of just touching the lady's hand, he gives the woman the impression that she is sexy and desirable. Finally, Wopat reminds us of the infamous male ego when he begs the woman not to hurt his pride. After all, what is the sense of that?

These lines even seem to have presidential pardon. The 2002 televised White House Christmas program included a rendition of "Baby It's Cold Outside." The words to the song, available at http://ntl.matrix.com.br/pfilho/html/lyrics/b/baby_it's_cold_outside.txt, provide a pickup line bonanza.

The Troy Donahues of today imitate the original. Donahue's signature red sweater, perfectly placed blond hair and sweet lines still seduce the ever-cautious Sandra Dee. How do women avoid these sneaky men who are only after one thing? Listen up, women, and "read between the [pickup] lines."

Chat Room

1. "Men's Pickup Lines" resulted from an activity designed to update the Donahue story. Compare/contrast it to the Donahue story. Discuss the different approaches taken in the two stories.
2. Compare/contrast a personality profile and a biography. Discuss similarities and differences.
3. Identify sentences that give context to the pickup line story.
4. Talk about the process behind the pickup line story. How do you think the story was conceived, researched and finally written by two authors?

Create

1. Take your birth year and subtract 100 years (i.e., 1982 – 100 = 1882). Profile a person who died in that year. When you have finished the profile, write a paragraph about the difficulty of profiling someone you couldn't interview.
2. Rewrite the lead and reorganize the pickup line story as you might if you were writing it for readers of a French magazine.
3. Write a definition for the word *clustering*, then write a paragraph about your own experiences with this technique or any other planning technique mentioned in the chapter.
4. Writing is a process. Think of other processes you know. List the steps in the writing process. Then list the steps for five other processes. List general similarities and key differences among them.

FEATURE STORY

"I Am an American: Ken Takaki and World War II"

Evelyn Gonzales
Author of "I Am an American: Ken Takaki and World War II."

Evelyn Gonzales, the 2003 recipient of the Laurie A. Lesniewski Award for creative writing, lives in Chicago. She frequently profiles her own family, exploring her rich ethnic roots. This feature, originally published in the Spring 2003 issue of The Avenue, demonstrates that ordinary, everyday people can be as engagingly profiled as movie stars.

"It was Dec. 7, 1941," he began as the glare of the fluorescent lighting hit his small face and big, round bifocal glasses.

Born in San Francisco, Japanese-American Ken Takaki sat in the living room of his modest two-story home on the north side of Chicago with a small smile across his tired face. All was still in the house as Takaki, 77, talked about his experiences as a Japanese male during World War II. The only other sound was his wife Eiko and younger brother Michael's muffled conversation in the kitchen.

Takaki's was one of many Japanese families living along the west coast of the United States before the war started. However, after the Japanese bombed Pearl Harbor, the United States evacuated all Japanese families, whether American born or not, and forced them into housing camps located further inland.

"They moved the families from their homes in Washington, Oregon and California to places like Idaho, Arkansas, Arizona and Utah," Takaki stated, pausing with each state name. "They didn't trust us; the government made us move."

Takaki, the first of eight U.S.-born siblings, was just a 16-year-old American when his family was forced to move in April 1942. President Franklin Delano Roosevelt signed an order forcing the evacuation in February of that year.

"The first place they sent us was to an assembly center. It was some kind of racetrack in the southern part of San Francisco," Takaki said. "It was the worst of 10 camps my family stayed at."

Far from their cozy home in California, the Takaki family did not have clean house stalls, and the smell of urine and stink constantly lingered in the air. Their mattresses were made out of hay and garbage bags.

"When we were evacuated, they told us only two handbags per person. We had to sell everything," Takaki said as he took a handful of peanuts his Filipino wife placed on the dining room table. "Merchants would come in and pay $20 for everything in the house."

After the family evacuated, they were given dog tags with identification and family numbers engraved on them, and any bag that couldn't fit on the train going to their camp was taken off in army trucks.

Getting food was no picnic either. In order to feed all the families at the camp, the army turned the field house into a big mess hall; the situation wasn't always pleasant.

"The first week we were there," Takaki laughed, "the ham was tainted. The Health Department had to be called in to help with the diarrhea. They didn't know how to feed us."

Although the food wasn't always what the families were accustomed to, "you got used to things." After a while the army stopped cooking creamy foods because "no one would eat them."

Once the army found out that Takaki's father was a cook, he helped prepare the meals for the entire camp. The army began asking other former cooks about what was best for the camp's population. "That's when we got rice instead of bread," Takaki nodded with a smile.

Their diets completely changed, families were provided with canned foods; meat and sugar were rationed. Yet life wasn't always bad.

"The government had talent shows to keep us occupied," said Takaki. That entertainment took the residents' minds off food for the time being.

However, the Takaki family did not spend long at this camp. In the subsequent year, they were relocated at Topaz, Utah. Although the war brought many changes, things did not change for the better for Japanese people, and especially not for his family.

The Takaki family was evacuated quickly and "didn't know enough to bring or buy winter clothing" with them to Utah. As Takaki spoke of the harsh winters, he rubbed his arms in reminiscence of the cold nights.

"It was a desolate area," Takaki said, recalling desert life. "There were desert storms, and when the wind would keel up, it kicks up dust and sand, and you can't face it. You have to turn your back." Takaki's eyes, now closed, showed the pain of what a desert storm feels like.

Yet weather conditions weren't the only problems Takaki had at his Utah camp. There was no privacy; everything was communal, including the showers.

"There were 45 blocks in our camp, and each block had 16 barracks," Takaki recalled. "In the center was a mess hall, laundry and latrine [bathroom]." He motioned to show the small dimensions of his living quarters on the dining room table.

Surrounding the blocks was barbed wire, which prevented camp residents from escaping. However, the barbed wire did not stop the military from mistreating the Japanese.

With a blank stare toward the wall, Takaki remembered an "older man in his 60s who got too close to the barbed wire. From the center of the camp, someone shot and killed him." Takaki's voice trailed off as he stared blankly toward the wall, leaving the story untold.

After this and other similar incidents, the Japanese were able to enjoy their constitutional freedom as American citizens. "After the [barbed] wire came down, we were allowed to hike and play baseball," Takaki smiled. When the American government gave them the freedom, once again, to enjoy baseball, he took advantage. He didn't know it would mean so much until he was able to play again. It's the simple things you miss, Takaki reminded.

Yes, America was giving back some freedom, but the Japanese were still kept under a watchful eye. Yet in spite of the living conditions, they were able to cultivate the land and raise farm animals for their own food supply. Takaki was even allowed to attend school, and he was the first graduation candidate in his camp. He graduated high school in May 1943.

Not really knowing what to do, in the latter part of that year Takaki decided to join the U.S. Army. However, because of his race and the American sentiment toward the Japanese, Takaki was forced to answer a questionnaire the Japanese-Americans called "No-no, Yes-yes."

"I can still remember two specific questions," Takaki said. "Questions 27 and 28 stated, 'Are you willing to fight for the United States of America?' and 'Will you denounce [the Japanese] Emperor?' "

Takaki answered the questions the only way he knew how, "yes" and "yes."

Because Takaki knew both languages, English and Japanese, the Army took advantage. They asked him if he wanted to help out with the war in the Pacific, and Takaki volunteered.

Therefore, Takaki attended a military language school in the Pacific to learn the lingo. Writing for the military proved to be difficult for Takaki. He was tested orally by the officers, whose "pronunciation wasn't that great." Takaki laughed at the thought of the American officers speaking broken Japanese.

Takaki was able to read and write Japanese, so the Army placed him in an advanced group. After six months of studying military language, Takaki passed the test and was "allowed to go into town." Now he would be able to enjoy his freedom.

Yet even though Takaki was an American soldier, Americans were still distrustful. "There were bed checks at night," Takaki remembered. He was still being watched.

Takaki's first army job was as an interpreter. He accompanied a Caucasian officer whenever needed, but this created problems because "some officers didn't feel we [the Japanese officers] were interpreting correctly," Takaki recalled angrily.

"When another officer questions you," Takaki said as his voice and eyes lowered to his lap, "you feel kind of bad."

He did not like to be considered un-American; his loyalties belonged to the United States, but people still distrusted him. And despite his life at the camp, Takaki was an American. He was an American during World War II and is an American today.

View

Writing a feature story involves the writing process: idea generation, research, organization of the material, drafts and revision, and the finished product. Throughout this process, the feature writer must keep in mind the intended audience: the reader. For example, a story for a local or regional publication must tie into the interests and background of its readers. The feature writer's idea, or angle, should relate not only to a particular time period, but also to its readers. The *profile* is a type of feature that provides relevant information to the intended audience about someone, something, some place or any combination of the three.

Help

cover story   the news or feature story pegged to the picture on a magazine cover. Other stories may be previewed in *teasers*, short captions meant to entice readers to look for these stories inside the issue.

formal outline   a written plan that requires the writer to put down ideas in full sentences and paragraphs.

global revision   the process of rewriting the story to clarify the writer's purpose and to convey the "gist" of what the writer wants to say.

informal outline   a written plan that does not require that details be expressed in complete sentences and thoughts.

lead story   the featured front-page story in a section of the newspaper, usually head-lined at the top of the page.

local tie   the sentence or paragraph that *ties* the story to the readership in stories written for a local or regional newspaper.

outline   a written plan or scheme for a story's main points and details.

personality profile   a feature story that delves into an individual's personality and tells the reader what that person is like and who he or she really is.

profile   a feature that focuses on (profiles) a person, place or thing and generally in-cludes background information, interviews and observed detail.

revision   a series of refinements as the writer works to perfect the story's structure and language.

secondary feature   a story slated for news-paper sections but intended to run as a sec-ondary rather than a lead story; a secondary feature may appear on front pages or inside sections.

trend feature   a story about a current fad, tendency, perhaps even a major social or political movement; topics may range any-where from fashion to terrorism.

Notes

1. William Zinsser, *On Writing Well*, 3rd ed. (New York: Harper & Row, 1985) 59.
2. Brendan Hennessy, *Writing Feature Articles*, 3rd ed. (Oxford: Focal Press, 1989) 43.
3. *Ad Age Almanac* 31 Dec. 2001:22.
4. *Advertising Age Fact Pack* 9 Sept. 2002:15.
5. Bruce Garrison, *Professional Feature Writing* (Hillsdale: Lawrence Erlbaum, 1989) 32.
6. Linda Flower, *Problem-Solving Strategies for Writing in College and Community* (New York: Harcourt Brace, 1998) 142–155.
7. Arthur Miller, "Introduction to the Collected Plays," *The Theatre Essays of Arthur Miller*, ed. Robert A. Martin (New York: Penguin, 1978) 141–43.
8. Flower 232–33.

3

The Style and Structure of the Feature Story

Style is the dress of thoughts.
                    —*Earl of Chesterfield*

To understand the relationship between style and structure, get out a few CDs. First, listen to the great tenor Luciano Pavarotti sing the Italian classic "O Sole Mio." Then spin a CD of the sometimes-pop tenor Andrea Bocelli singing the same song. Next, for contrast, play Elvis Presley's version, "Now or Never." As far as the music—the notes—you have the same melody, the same song. We might say the structure remains pretty much the same from one track to another. But, ah, the voices. That's another matter. The word *style* derives from the root word *stylus*, to engrave your mark. Each singer imprints the structure of "O Sole Mio" with his own style.

Levels of formality differ among the three singers, with Pavarotti's style the most formal and Presley's style the least formal. The level of formality relates directly to the technique: Pavarotti's extensive formal training gives him a polish, a surface perfection, that the untutored Presley's voice never acquired. Bocelli falls in between, though his formal training with Italian tenors puts his style much closer to Pavarotti than Presley. Nevertheless, it could be argued that Bocelli's passion falls closer to Presley's raw emotion. Pavarotti's style might be equated with satin or silk, smooth and elegant. Presley's unpredictability and fuzzy warmth could be likened to the teddy bear of which he sings. Bocelli crosses styles, opera with pop. In an interview for the video "A Night in Tuscany," he admits that opera and pop are "two very different things." But he can't say that one is superior or inferior. *Vive la difference!*

Even the words of the various renditions show stylistic distinctions. "O Sole Mio" reflects the Petrarchan sonnet tradition with its high conceit ("Oh my sun!") while Presley's "It's Now or Never" uses the American vernacular. "O Sole Mio" metaphorically exalts the beloved as higher and brighter than any worldly creature, as ethereal and unobtainable as the sun. Presley's adaptation brings the woman down to the literal level; he implores his "darling" to kiss him, to be obtainable. His now-or-never urgency reverses the conceit of woman as unobtainable sun and insists on consummation "tonight."[1] Presley's American lover has no time for prolonged Petrarchan courtship; tomorrow would be "too late" for very real arms, lips and love.

Style frosts the cake of life: From a funky outfit to casual behavior to gospel crossed over to the blues, style makes life interesting. Structure holds life together.

And so it is with the rainbow world of the feature. We can learn set principles of structure, but style spans infinite differences from writer to writer and publication to publication.

Feature Structure

The purpose of the news story is to inform. Thus, news writing traditionally takes the shape of an inverted pyramid, with key facts quickly and cleanly presented in the lead and additional, supporting information presented in declining order of importance. Unlike the news story, traditionally written in inverted pyramid style, the feature story takes the shape of an essay: it has a distinct beginning (feature lead), middle (body paragraphs offering support for the writer's focus/point of view) and conclusion (often a return to the image or idea introduced in the lead). The feature frequently is structured in the narrative mode, telling a story in chronological order.

While writing the lead is one of the most difficult aspects of feature writing, crafting a lead that will draw the reader into the typically longer feature story constitutes one of the form's most creative challenges. The lead must interest the reader yet remain consistent with the story's central idea. Take, for instance, the lead for the Colm Feore profile that follows in this chapter:

> Looking comfortable in a T-shirt and jeans, Canadian actor Colm Feore arrived at the appointed time in the Stratford Festival's VIP Room. He carried a pack of cigarettes, ready to chat. He is a man, he says, who has committed his life to language.

This *descriptive lead* gives color to the profile of the actor; it paints a picture of the man as the interview began. It also establishes character, much as the first paragraph of a short story might. Suppose this piece had been written as a hard news story. The inverted pyramid lead might read:

> Colm Feore, dubbed the "rising star" of the Stratford Festival, opened as the title character in Shakespeare's "Richard III" last night in Stratford, Ontario. He is also starring in two other mainstage productions at the Festival Theatre, "The Three Musketeers" and "The Taming of the Shrew."

See the difference? The news lead stays strictly with facts and addresses all 5 Ws (who: Colm Feore, where: Stratford Festival in Stratford, Ontario, when: last night, what: stars in three mainstage productions, why: he's the Festival Theatre's rising star). The feature writer, in contrast, may choose from a variety of ways to "hook" the reader into the story. The feature lead must attract the reader responsibly, of course. The reader should not be misled by the lead or jarred by the unfolding story content. Jon Franklin refers to the great Russian playwright Anton Chekhov in phrasing this principle as Chekhov's Law: "If the opening of a story mentions a shotgun hanging over the mantel, then that shotgun must be fired before the story ends."[2]

The Lead: How Do I Begin?

In Chapter 1 you were introduced to the *suspended-interest lead* in Bill Moor's news story. (If you don't remember the definition for this type of lead, do not "pass go." Return to Chapter 1.) While suspending interest is one way to begin a feature as well as a news story, use this type of lead, as well as the *descriptive lead* (just mentioned), sparingly. Other types of leads exist and should be chosen for their appropriateness to the story's central idea or *angle* (see Chapter 1, Click Here, on p. 6 if you missed this definition).

Since the community remained puzzled about Charles Feirrell's behavior and subsequent legal fate, Moor wisely chose the suspended-interest lead to pick up on the prevailing mystery surrounding the story. Leaf back to the first page of Chapter 1 and notice that Moor's lead offers the reader plenty of white space. Consider how you feel about copy-heavy textbooks. Few readers find big blocks of type an enticement to continue reading. The lead should be as brief and punchy as possible unless, of course, you're writing for The New Yorker or another magazine whose style tends toward hefty leads.

To get the reader's interest fired up, you might describe the person to be featured in the story, as the Feore lead does, or you might set the stage by describing the setting. The type of feature known as the personality profile often benefits from a descriptive lead. Depending on the type of feature you're writing and your angle, you might begin other ways. Decades ago, Carl H. Giles identified the basic types of feature leads: impact statement, description, narration (anecdote), contrast, direct address, quotation, and question.[3] Let's take a look at some of these types.

Impact Statement

You've seen the lead that begins with a shocking statement or statistic. The writer intends this initial jolt to shock the reader out of complacency, to "surrender his attention to the statement."[4] Sometimes the impact statement doesn't "throb with shock" but is "nothing more than a pure human-interest fact." For example, a story I wrote about Tom Wopat—who, in fact, starred in the 1980s television series "The Dukes of Hazzard" and has since headlined on Broadway—began with such an impact statement: "Tom Wopat resists the idea that he is a star." Hopefully, readers wished to continue reading to discover why someone who had achieved star status would not want to consider himself a "star." The answer to this initial question became the angle of the feature.

Description

The descriptive lead need not be reserved for the personality profile. Anna Quindlen chose to use description to set the stage for her feature published in the special September 11 double issue of Newsweek magazine:

> Nightfall is as dramatic as the city itself in the days surrounding the winter solstice. The gray comes down fast, pearl to iron to charcoal in a matter of minutes, muting the hard edges of the buildings until in the end they seem to disappear, to be replaced by floating rectangles of lantern yellow and silvered white. In the space of an hour the city turns from edge to glow, steel to light.[5]

This description sets the stage and, thus, the nostalgic tone for Quindlen's opinion piece, a personal reflection on the aftermath of the Sept. 11 attacks on the World Trade Center. Her language choices include words we have come to associate with the tragedy, such as "steel" and "hard edges," common descriptors of what is now known as Ground Zero. The colors described both locate the reader in the cold city and evoke the feeling of sadness associated with gray color. Her observation that the buildings of New York City "disappear" in the fading light especially chills readers who are unspeakably aware that, in fact, the World Trade Center essentially disappeared. The feature that follows conveys her angle: As the city's skyline has changed, so have its people been "changed forever by grief."

Narration

A feature may begin with narration, a brief *anecdote* (a short account of an interesting or humorous incident). Liz Krieger's feature on eating disorders begins with a definitely not funny account of a woman's frustration shopping for large-size clothing:

> Filled with wonder and delight, she walks into the clothing store. So many colors to choose from, so many fabrics, and so many styles! She grabs the first black silk blouse she sees and heads for the dressing room, but first makes a check of the tag. It's a Small size. She goes back to the rack to get a larger one. She checks thoroughly, but only sees long rows of Smalls with a few Mediums peppered throughout. No Larges, and certainly no Extra Larges, are anywhere to be found. Her mouth moves into a frown, and she walks out of the store with no silk shirt, her head hung low.

The story should strike a chord of recognition for many readers who have had similar experiences. As it creates reader identification, this anecdotal lead also introduces the feature's focus on the frustrations of life in a society oriented toward thin women. Liz's lead accomplishes what Giles says the narrative hook should. It attracts the reader with a "dash of drama" and "capitalizes on the urge everyone has to hear a good tale" or encounter "a slice of life."[6] Whether the tale is serious or humorous depends upon the story that follows.

Contrast

The presentation of two different but feasibly comparable places, persons, or things in the lead "helps to identify something by associating it with an extreme or an opposite."[7] For example, in an article originally published in Vanity Fair, Dominick Dunne stresses the horror of Nicole Simpson's murder when he compares her to another Mrs. Simpson in his contrast lead:

> For sixty years, whenever the name "Mrs. Simpson" was mentioned, it belonged, irrevocably, to Wallis Warfield Simpson, the lady from Baltimore for whom King Edward VIII gave up his throne in 1936 and shook the British monarchy to its roots. No longer. Now the name belongs, irrevocably, to the tragic and beautiful Nicole Brown Simpson of Brentwood, California, whose dreadful death on the night of Sunday, June 12, 1994, along with that of her friend Ron Goldman, has riveted this country for nine months. The nastiness of how she died and where she fell is the reason the gaze of the country is focused on a courtroom in downtown Los Angeles.[8]

Direct Address

When the lead uses the second-person pronoun *you*, directly addressing the reader as though in conversation, the informality bridges the writer-reader gap to create a sense of intimacy and identification. The writer invites the reader inside the story. Molly V. Strzelecki begins a feature by asking the reader to imagine how it feels to have writer's block, the subject of her story:

> A large, muscular linebacker barrels at you and lunges for your notebook as you race toward the end zone. A goalie, drowning in padding, takes a keyboard to the face mask as you take your best slap shot at him. A six-foot, ten-inch point guard easily knocks down your pen as you shoot for the basket.

Quotation

Starting with a quotation may seem an easy way to open a feature, but only a truly arresting, well-phrased quote holds up as a lead. If the content creates interest but the quote isn't beautifully phrased, then paraphrase. Avoid lengthy quotes. Identify speakers unless, of course, as with the following example, the quote is famous enough to be easily recognized:

> "Good evening. Today, our fellow citizens, our way of life, our very freedom came under attack in a series of deliberate and deadly terrorist acts."[9]

Kaitlin E. Duda chose to begin her feature "A Society Challenged and Changed/The Day I Grew Up" by quoting President Bush's 9-11 speech. She pulled partial quotations from the speech into the structure of her feature, employing his points as counterpoint for her own words and ideas. As she wrote her feature, Kaitlin interwove Bush's words into her own narrative.

Question

Students find the question lead another simple way to start a feature. For that reason, I was surprised to have such a hard time finding one as I flipped through numerous magazines. In fact, the long leads in Newsweek, Mother Jones and Modern Maturity also surprised me. I turned to a newspaper, where I expected the leads to at least be briefer. A secondary feature in the local newspaper turned out to have the only question lead I could find. The question generally poses a problem; the solution emerges from the feature. The following lead asks a question and immediately answers it:

> Mix sunlight, temperatures near 40 degrees and melting snow and what do you have? Well, if it's the first week of January, a beautiful day for a car wash.[10]

Here the question lead attempts to capture reader interest by presenting a puzzle rather than a problem.

How Do I Continue?

A great lead engages the reader in the story, but even the strongest lead is unlikely to give focus to the feature.[11] The *nut graph*, which conveys the story's thesis (main idea), "usually comes right after the lead is established, or played out." The nut graph may go by other names—*bridge*, *focus statement* or the spelled-out *nut paragraph*, but no matter what it's called, the short-and-sweet nut graph is an essential component of a feature story.

As an example, remember the Troy Donahue feature in Chapter 2? The two-paragraph lead presents an irony, but not the main idea of the story:

> Like most teenage boys in the 1950s, Troy Donahue admired the artistry of two rebels with talent—film actors Marlon Brando and James Dean.
>
> Unlike most teenage boys, Donahue ventured to Hollywood to pursue the paths of his heroes. He had just graduated from New York Military Academy.

The next paragraph sets up the main idea of the feature:

> Instead of becoming the next Brando or Dean, Donahue became another kind of American icon, which he described in a recent telephone interview as the "innocuous, all-American goodie-two-shoes guy."

This paragraph *bridges* the initial irony to a second irony, then to the idea that the story will pursue—how and why Donahue became an American icon. This paragraph is as essential to the feature as an eye-catching lead. Furthermore, this nut graph also provides *context*—the reader now understands that the interview took place via a recent telephone call.

Once the nut graph has been written and context has been established, feature writers must find a way to transition into the body of the feature. Writers draw from transitional words and phrases such as these:

- *chronological transitions:* then, next, afterwards, later, moments later, finally, meanwhile, in the meantime, at last, eventually
- *comparative transitions:* accordingly, likewise, in addition, similarly
- *contrast transitions:* however, even so, even though, instead, on the other hand, yet, whereas, still, nevertheless, on the contrary
- *transitions that point out the obvious:* unquestionably, undoubtedly, indeed, admittedly, granted, no doubt, certainly, no one denies, of course, while it is true that
- *transitions that set up examples:* for example, for instance
- *transitions that show reaction and cause:* as a result, consequently, thus, therefore, logically

Sophisticated writers find other ways to create transitions, such as repeating key words, phrases or images, or bridging paragraphs with synonymous words and phrases. Narrative bridges allow one scene to build upon another and facilitate flow. For example, the one-sentence paragraph, "Perhaps he [Feore] should think of himself as an athlete," bridges the idea that the actor refuses to see himself as a star but insists on doing his own stunts in the Feore story that follows.

In general, organize the feature according to the material you have to present. Feature writers collect factual information about their topics and create notes as they interview and observe. As discussed in Chapter 2, this material may then be grouped or clustered into subtopics before the story takes shape. Some stories may be organized chronologically, spatially or causally. Some material may lend itself to comparison and contrast. A concession or definition may be required near the beginning.

Recall the *rhetorical modes* you learned in composition classes. Feature writers use the same methods as essayists to structure paragraphs and stories: narration, description, cause and effect analysis, compare/contrast, classification and division, and definition. In addition to the narrative mode, descriptive writing, which takes spatial organization, appears extensively in feature writing. Writers selectively describe people, places, actions and events from their own observations. The rhetorical mode determines how information will be sequenced; if they have been successfully sequenced, images and words will unfold smoothly in the story.

The feature should be structured with logical chronology. Whether the writer chooses to tell the story in past or present tense, the verb tense should be consistent. The reader won't appreciate being yanked around by tense shifts. Occasionally, the writer may shift tense for a logical reason; for example, the writer might use a *flashback* to provide background information that occurred before the time of the story or a *flashforward* to project what is expected to happen in the future.

Foreshadowing requires that the writer take a part of the story out of sequence in order to "insert details in the story that will allow him to conduct his dramatic scenes without the necessity of explaining background details" later.[12] Flashbacks, flashforwards and foreshadowing can be valuable storytelling techniques, but they do manipulate time and somewhat disrupt the flow of the story; thus, they should be used sparingly.

The time element of the feature also relates to the story's timeliness. The story must relate in some way to the time of its publication. For example, a historical feature about the underground railroad has greater news value if it is linked to Black History Month and published in February. Magazine features may have tie-in paragraphs that link stories to themes of specific issues. As mentioned in Chapter 2, a *local tie* may be required, especially for newspaper features. For example, readers of The Kansas City Star might not be interested in a feature about national observations of Black History Month unless a paragraph ties these observations to local activities.

Feature writers use the structures of the rhetorical modes to cast their stories into the formats distinctive to feature writing. These formats include the personality profile, the historical feature, the Q and A (in which questions asked and answers given during an interview are presented sequentially), the sidebar (a short, supplementary boxed item that provides background or additional information supportive of the main story), the opinion piece (which directly states the writer's point of view), the personal experience feature (which takes first person voice), the humor and advice column, specialty writing (food and fashion features), color features, sports and travel features, and entertainment reviews.

How Do I Stop?

Feature writers can take cues for great conclusions from fiction writers. Think of your favorite novels and short stories. How do they end? Margaret Mitchell's "Gone With the Wind" haunted me for years because it posed an open ending. Scarlett O'Hara has lost Rhett Butler; how will she get him back? *Will* she get him back? Mitchell leaves the reader with only Scarlett's hopeful statement, "I'll think about it tomorrow. After all, tomorrow is another day." At age 12 I was so intent on a satisfactory ending that I penned my own in my parents' copy of the book. This brought me nothing more than a lecture about not writing in books.

An inconclusive ending may leave readers thinking about the problem indefinitely. Other endings can be more satisfying, but the best endings are always suggestive. You're not writing mystery stories in which all the loose ends must be tied together by some *deus ex machina*, some sudden revelation. The feature should reveal insight into a situation, character or event as the story unfolds; it should not just be tacked on at the end. As you read the sample essays in this book, works of fiction and nonfiction for your classes, and magazine and newspaper articles at your leisure, study the concluding paragraphs. Which do you prefer? What do they do? How do they do it? Writers learn from reading other writers and asking these questions.

Publication Style

Each newspaper and magazine has its own style. If you wish to write for a specific publication, you need to study it first very carefully. How are its features structured? What topics are appropriate to its readership? Finally, what is the publication's style?

To succeed as a feature writer, whether you write for newspaper, magazine or public relations purposes, you will need to learn basic rules of style that vary little from one publication to another. Individual publishers may have stylebooks of their own, but all share most of the rules referenced in the stylebooks published by two major news service agencies, the Associated Press (AP) and the United Press International (UPI). The Associated Press Stylebook defines its organization as a "newsgathering cooperative dating from 1848." The purpose of its Stylebook is to make "clear and simple rules, permit few exceptions to the rules, and rely heavily on the chosen dictionary as the arbiter of conflicts."[13] The AP Stylebook has expanded from 52 pages in the early 1970s to almost 400 pages in recent editions. The UPI, a 1958 merger of the United Press and International News Services, publishes its own stylebook.

As you read the following feature stories, try to determine what style decisions had to be made. The features demonstrate the characteristics of feature structure: *transitions* (the connective tissue that allows the text to flow from sentence to sentence, paragraph to paragraph), *unity* (a central focus, problem or question), descriptive details and quotes. Look for these characteristics as well as for the elements that create emotion, human interest and personal style. Finally, look for what Hemingway claimed was missing in news writing: the why—why certain events occur or why people do what they do.

Personal Style

Within the guidelines of publication style, writers learn to project their own voices. As explained at the beginning of the chapter, style may be formal or informal; however, the style should always be appropriate to the topic. For example, as Winifred Bryant Horner says in "Rhetoric in the Classical Tradition," words such as *ain't* or *cute* would be as inappropriate for a story on nuclear disarmament as a tuxedo would be wrong for a picnic.[14] Horner tells us that the writer's point of view, the topics, and the audience and occasion determine the writer's stylistic choices. Specifically, writers might convey color and tone through some of the following stylistic choices:[15]

- colloquialisms: informal phrases common to spoken language, including contractions ("he'd rather walk a mile").
- localisms: words and phrases that vary according to geographical areas (such as regional differences in referring to soft drinks as "pop" or "soda").
- slang: informal expressions common to a group of people which serve to bond that group (a "slacker" "kicks back" with his friends while an official receives "kickbacks" from "shady" deals).

- jargon: formal technical terminology shared and understood by a specific group of people or members of a profession (physicists might talk about "rads," but I usually don't).

In addition to word choice, the structure of sentences and paragraphs reflects personal style. One of the distinctive styles to emerge in the 20th century was that of Ernest Hemingway. His expertise with the simple declarative sentence, which he credited to his early work as a journalist, came to characterize the lean, masculine prose that made him famous. His writing style and subject matter reflected his larger-than-life personality; he was a man of action, a world traveler who hunted big game in Africa, immortalized the bullfights of Spain and witnessed the horrors of trench warfare. A writer's style should reflect the person; that is what we mean when we say a writer has found his or her "voice."

Writers and other artists have apprenticed with the masters for centuries. As Picasso imitated the masters who preceded him in his early education, so writers have translated the great works from other languages and imitated the great stylists through writing exercises. It has become a cliché that imitation is the highest form of flattery. As mentioned in Chapter 1, my class admired Molly Donnellon's "Making Stone Soup" so much that they imitated her.

As a student in an undergraduate creative writing class, I voraciously read the works of F. Scott Fitzgerald. My professor didn't like it very much when I began to sound like a bad imitation of his voice. She said, "Why don't you read Hemingway? He's cleaner." My first published short story came from my "Hemingway phase." I now "prescribe" Hemingway for my own students. The imitation that concludes this chapter was written by "prescription." Jennifer Jones, the student who wrote it, felt it was the best story she had ever written. A faculty member who evaluated her writing portfolio for the college writing requirement agreed. She had taught Jennifer during a previous semester, and her response to "A Moveable Moscow" was, "I didn't know Jennifer could write like that!" You can't stop Jennifer from writing now.

My feature writing class was asked to consider the writing style in the Feore feature. We had already listened to a CD of Elvis Presley's "It's Now or Never," watched a concert video in which Andrea Bocelli sang "O Sole Mio," and studied style in a Britney Spears concert video and an Enrique Iglesias music video. The students were then sent off to the computer lab to rewrite the Feore feature conclusion and translate the musical style of one of these artists into writing style. Sara Pendley, Renée Donovan and Evelyn Gonzales gave long, Latinate sentences and sensual description to their Latino-style conclusion:

> In the open doorway, his masculine silhouette was defined by shadows from the crisp moonlight of a brisk Canadian evening. A long drag from the glowing red Marlboro, smoke escapes his satisfying lips, as he ponders the questions, "Where would all this energy take him? Would he return to the Festival Theater in 1989?" He raises his brows, revealing his deep-set, engaging eyes while quoting from "Othello" the eloquent words: "I've done the state service and they know it." His answer reveals he is more than a fine figure: He is a true Don Juan.

✎ *Click Here: Stylus—Engraving Your Mark*

Those pesky (or pestilent) pronouns!

This is the question I hear more than any other in my writing classes: "Can we use the word *I*? My high school teacher said to never, never, never, never, never use it!" The question usually comes up during peer critique sessions as a peer critic encounters the dreaded pronoun in another student's draft. The pronoun *you* has also been known to provoke verbal fisticuffs.

A few semesters ago, I agreed with the high school teachers. For a whole semester, I battled students who just couldn't seem to write a feature without a personal pronoun. Then I read "Literary Journalism," edited by Norman Sims and Mark Kramer. After I read the second essay, Kramer's "Breakable Rules for Literary Journalists," I realized that my students and I grew up with very different schools of journalism churning out the copy we regularly read.

I grew up with what might now be called the "old" journalism (in contrast to the "New Journalism" of the mid-1960s introduced by writers such as Tom Wolfe). This period of inventiveness "broadened the form" and led the way for journalists in the 1980s "to identify themselves as part of a movement."[16] Characteristics of the new, literary journalism and its practitioners include hanging out with "sources for months and even years," cutting corners on accuracy and candor—recasting and even constructing events, and using "intimate voice" in contrast to the impersonal voice of the "old" school (otherwise known as objective third-person *he, she,* and *it*).[17] A style decked out with "efficient, individual, informal language" distinguishes the literary journalist.[18] Think of it as the triumph of Presley over Pavarotti.

I learned my lesson and gave up the fight—with some qualifiers, anyway. I never, never, never, never, never allow first or second person (*I* or *you*) in personality profiles. After all, emphasis should be on the person profiled, not on the writer or reader. I also caution students to study a publication's style before they write for it. Some publications still uphold the "old" practice of objective journalism. If you want to publish with them, your style must conform.

Some students, who have found that the formal style of academic writing just does not fit their personalities, have watched their prose take on new life in a feature writing class. More than any other type of nonimaginative college writing, the feature allows personal style to emerge.

FEATURE STORY

"Colm Feore"

A time limit of one hour was set by the media director who arranged this interview with actor Colm Feore. However, from the start of the interview it was clear that this man loved to and wanted to talk. It was an easy interview—he chatted freely and, as though words weren't enough, added action to illustrate the points he made. The hard part, initially, was getting away. Later, it was sorting out who this man really was—he was highly fit but a wiry chain smoker, a man of action yet a man of words.

Feore has apparently left the stage for a film career—he no longer appears at the annual theatre festival in Stratford, Ontario. Since this interview was published on Aug. 7, 1988, Feore has appeared in over 40 films, including the television serializations of "Storm of the Century" (1999) and "Nuremberg" (2000), and in feature films such as "The Caveman's Valentine" (2001). In 2003, along with other members of the cast of the movie musical "Chicago," he won a Screen Actors Guild Award for Best Ensemble Acting for his portrayal of Assistant District Attorney Harrison.

Looking comfortable in a T-shirt and jeans, Canadian actor Colm Feore arrived at the appointed time in the Stratford Festival's VIP Room. He carried a pack of cigarettes, ready to chat. He is a man, he says, who has committed his life to language.

Certainly, he likes to talk. Feore, referred to earlier that day by the Toronto Star as the Festival Theatre's "rising star," canceled an audition in Toronto in favor of the interview.

"I was going to try to do both," he explained, "but my wife said it would be too much." So he chose the interview. "So many people come [to Stratford] from your area," he added, referring to the southwestern Michigan/northern Indiana/Chicago area with its proliferation of group tours to the Stratford Festival.

Stratford, Ontario, first hosted its annual Shakespeare Festival in 1953. On July 13 Sir Alec Guinness, as the title character in "Richard III," walked onto a stage erected beneath the second largest canvas tent in North America. The 2,300-seat Festival Theatre, famous for its thrust stage, replaced the tent in 1957. Eventually, Stratford's old opera house became the Avon Theatre, and a new theater became the Third Stage. From May until November each year, the festival draws actors and audiences from all over the world.

Feore currently plays lead roles in three of the Festival Theatre's productions. He had just opened the night before as Athos, one of "The Three Musketeers," a performance he would repeat again that night, bathed in golden lighting, wearing dark, earthy costumes in lush fabrics, and executing a fight scene in which he throws off three men at once. The next day he was scheduled to play a matinee in the title role in William Shakespeare's "Richard III," and to become Petruchio in "The Taming of the Shrew" by night. Feore himself must wonder sometimes who he is.

He was looking leaner than he had in the 1987 season when he received rave reviews for his portrayal of Iago in the Festival Theatre's "Othello." A tough schedule? "Sure it is," he said.

He paused before lighting the first cigarette, taken back a moment by the designation "rising star." After all, Stratford's ensemble company has resisted the star system over the years.

Blowing smoke, Feore balked at the tag. He didn't see any "rising" in his Stratford career, which began eight years ago when he walked onto the stage, an apprentice actor cast as Romeo. That, he says, "was a risk. This season is not risky." He prefers to think of himself as a "reliable leading man" rather than as a "star."

Perhaps he should think of himself as an athlete.

This season Feore carries the most physically demanding roles in the company. "The Three Musketeers" requires skillful fencing and fight scenes, "The Taming of the Shrew" demands youthful energy, and "Richard III" inspires him to difficult and dangerous feats.

Many actors let a limp suffice to display Richard III's mental and physical distortions. Others use mechanical devices to throw their bodies into distorted curves.

"I don't use a steel brace as some actors have," Feore said. To illustrate, he stood up and twisted his hips in two impossible directions, contorting his arm and hand. "I stay like this for two hours and have to move quickly. Richard moves faster than anyone else because his handicap forces him to live and fight in different ways from others."

For the climactic fight scene between Richard and the Earl of Richmond (Geraint Wyn Davies), Feore refused to work with the customary fiberglass weapons and shields. "Did you ever hear a sword hit a fiberglass shield?" he asked, then mimicked the sound. "I insist on realistic steel weapons."

Consequently, he performs the fight scene covered with spiked steel armor, carrying a 15- to 20-pound shield, wielding a seven-pound steel hammer. "It's hack and hew," he said.

Armed as the actors are, the scene is definitely risky. He recalled that when his wife attended the play's premiere, she couldn't watch the fight scenes. "She hid her eyes, but she could still hear the clanking and the heavy breathing and the cries." He and Davies literally and "regularly get beaten to hell" as "steel sparks fly," he said, apparently quite excited by the idea.

In "The Three Musketeers," the right-handed Feore is challenged by a scene in which he fences with his left hand, his right hand in a sling.

Audiences may not realize that stage fighting is more "real" than fake. It requires special training. Most of Feore's fight training has been with the Festival Theatre's associate fight director, Patrick Crean. Once a fight and stunt director and "double" for Errol Flynn, Crean has directed countless spectacular sword fight sequences for stage, film and television.

Crean brings some of the old film swashbuckler techniques to "The Three Musketeers," with Feore executing a "Hollywood parry" in which he kills someone in three moves. "That's a real surprise for the audience," he said.

He must maintain this level of physicality in his portrayal of Petruchio as he comes to blows—not with a man—but with his wife Kate. They hurl plates and utter epithets in Italian (Shakespeare's play is set in Italy, after all). Feore must defend this blatant display of domestic violence. After all, this is the twentieth century, not the sixteenth. Those who see Petruchio as chauvinistic have "a limited mindset," he said in his character's defense. "Petruchio is only parroting Kate. He registers everything she says and gives it back to her."

The current production redresses the Renaissance comedy in 1950s style. Feore rides a motorcycle and wears a black leather jacket and tight jeans, a Marlon Brando/James Dean antihero, a rebel with a cause—in this case, to tame his Kate. He co-stars with Goldie Semple who portrays Kate as a flaming rebel who flashes her red petticoats. The '50s motif blends with the Italian in even the minor details: a car radio plays "The Great Pretender," linking '50s music with Shakespeare's theme of disguises; Lucentio drives onstage in a red Italian sportscar with Pisa license plates; and the nerdish Hortensio wears white socks.

The appointed hour had expired, but Feore lit another cigarette, leaned back, and began to talk about his Irish origins. Born in Bost, he spent "a few formative years" living in Ireland before the family moved to Windsor, Ontario. It was inevitable that he would end up in the theater. The marriage between the Irish and theater comes, he said, "from the oral tradition," and, with a grin, he added, "from storytelling, lying, cheating and dissembling." The words rolled off his tongue.

Acting on the Stratford stage doesn't pay the bills, though. "Commercials pay the bills. It's absolutely terrific!" Feore has decided to buy a house in Stratford, a convenient two-hour commute to Toronto where he freelances for Canadian television commercials. Since Canadian law prohibits the use of non-Canadian actors in commercials, his voice regularly replaces the voiceovers of American actors such as Eli Wallach and Martin Sheen.

Without such income, many actors and directors "could not afford to live in a shoe box," he said as he carried the conversation all the way to the theater's outside doors. Even though he would be on stage again in just two hours, he was vigorously involved in the conversation, an irrepressible bolt of energy in blue jeans. Would he never get tired?

Chain-smoking, hanging in the doorway with yet another cigarette, he looked more like a ripped-muscled biker than the swashbuckler-of-the-night. Where would all this energy take him? Would he return to the Festival Theatre in 1989? His answer was a quote from Othello: "I've done the state service, and they know it." No monosyllabic biker, this.

Chat Room

1. This feature educates as well as entertains. What does this one teach about acting?
2. Features require accurate reporting. Identify facts and details that show attention to accuracy.
3. The successful feature writer has a strong sense of curiosity. This is often evident when the story asks a unifying central question that may or may not be answered by the end. Does this feature ask a central question? If so, is it answered?
4. Identify the nut graph, transitions and quotes. Discuss the role these components play.

Create

1. Description is a distinguishing characteristic of the feature. Visit a professor's office. Observe and record details—furnishings, furniture arrangement and color scheme. Describe the professor's appearance and actions. Can you draw an insight about the person from your notes?
2. In preparation to interview actor Colm Feore, I attended "The Taming of the Shrew" and "The Three Musketeers" to observe his performance and add descriptive color to the feature. Observe someone involved in an activity, then interview the person.
3. Consult several short story collections. Copy down the beginning paragraphs of three or four of these stories. Write a short analysis of the purpose or effect of each introduction.
4. Copy down the final paragraphs of three or four short stories. Write a short analysis detailing how each conclusion affects you.

"After the Cheering Stops"

Eric Hansen
Author of "After the Cheering Stops"

Writing features for sports requires a muscular style and active voice to reflect the vigor of the topic. Even sportswriters must deal with intangibles, though, and very often with defeat as well as victory. Eric Hansen, South Bend Tribune staff writer and managing editor of the national news magazine The Irish Sports Report,[19] commented on writing the following feature: "One of my favorite themes in stories is dreams, not the ones that follow the perfect script with smiles paving every step along the way, but the ones that take a detour. Where do dreams go to die? And how do we bring them back?" This feature reminds us that not all persons profiled need to be celebrities or even winners.

It was never her own dream, nothing Dayna Pellosma ever had written on a list and tucked away in the back of her heart.

Running track in college was something seemingly everyone else wanted her to do, felt she was destined to do. Eventually, that's what made running not fun anymore.

The expectations. The suddenly serious faces on her coaches. The stacks of letters from Harvard, Purdue, Auburn and many colleges the former LaVille, Ind., sprint standout had never heard of.

It was more than overwhelming.

It was hell.

So during her junior year in high school, a week into track season, the defending Indiana state 100-meter champion simply walked away, leaving behind two full years of high school eligibility and a lifetime of promise.

"We really didn't see it coming," said LaVille assistant coach Paul Papczynski whose daughter, Lori, was a teammate of Pellosma. "And I guess we really didn't understand why.

"It's a shame, because she had everything laid out for her—more recognition, more championships, a college scholarship wherever she wanted to go. But we weren't going to kiss her butt to bring her back."

And so the dream died there . . . or so it seemed.

Today Dayna Pellosma is living in the Smoky Mountains of North Carolina. The 24-year-old has dropped the *y* from her first name and hyphenated her last name, the latter a product of marrying a man named Jason Kelly, whom she met in North Carolina.

"He's in the military and he's stationed in Japan," Pellosma-Kelly said. "He's going to be there for another two years, so we only really get to see each other about every six months. It's tough. It's kind of crazy."

Yet almost every other aspect of her life is grounded in pragmatism.

She attends a junior college near her home in Dillsboro, N.C., a quaint tourist trap with about 400 people and one traffic light. For excitement, she can head to Pumpkintown, Birdtown, Love Field or Unahala.

"If you really want excitement, there's always Sylva," the Jackson County seat, population 5,000.

But North Carolina is home now and a perfect place to restart the dream.

Her own dream.

Pellosma-Kelly didn't have a track scholarship to fund her education at Southwestern Community College, so she works part-time at a gas station and receives grants for her consistently good grades.

"I used to feel so bad about not taking those track scholarships, because my mom worked so hard to give me a good life," Pellosma-Kelly said. "I felt like I had everything handed to me on a silver platter and I threw it all away.

"But this means a lot to me, that I've earned my way through school through hard work and good grades. I know I've made mistakes, but I'm kind of happy with the way things are turning out. This is a good life."

And it may be on the verge of getting better.

Pellosma-Kelly had contacted Western Carolina University, a nearby Div. I school in Culowhee, N.C., about resuming her track career. However, she never reached coach Danny Williamson. Instead she was erroneously routed to a secretary, who told her she was too old.

"It kind of bummed me out," she said. "I haven't run competitively since my sophomore year in high school, but I've kept in shape through hiking and mountain biking. I know I can do it. All I want is a second chance."

That may be coming. Through a quirk of fate, Williamson found out about Pellosma-Kelly, about her rise from obscurity to dethrone defending champion Angela Young of perennial state power Gary Wallace at the 1993 Indiana state meet. And Williamson plans to meet with her in the coming weeks and invite her to join the team if all the details can be worked out.

"She might be exactly what we're looking for," Williamson said. After some short-lived skepticism, LaVille coach Lee Dennie and assistant Papczynski sure felt that way the first day Pellosma showed up for track back in the spring of 1991.

When she settled into the blocks the first time, Pellosma whinnied like a horse.

"That kind of became her trademark," Papczynski said. "She even ran like a horse. We thought about trying to change her stride. But when we looked at the stopwatch, we decided to leave her alone. She was something special."

LaVille had already established itself as a something special among the state's small schools. Pellosma added to the tradition by finishing seventh in the 100 at the state meet as a freshman.

"The only struggle was getting her to practice," Papczynski said. "Her mother was a single mom and couldn't take her to and from practices and meets. So we made those arrangements.

"She also couldn't afford her own track spikes. So one day, we went out and bought her some. But we told her we found them, so she wouldn't be embarrassed. She cried she was so happy. It seemed like the beginning of a fairy tale."

That story line continued Pellosma-Kelly's sophomore season. Despite not having the benefit of racing against any of the state's top sprinters during the season, Pellosma-Kelly stormed to the 100 title, defeating defending state champ Young and posting a time of 11.97.

"I was so overwhelmed as a freshman, but I was confident this time," Pellosma-Kelly said. "Plus, my mom was in the stands this time, so I felt compelled to win."

The euphoria gave way to anxiety during her junior season.

"I think the coaches got a little too serious with her," teammate Jen Crone recalled. "And I think we just took for granted how good she was. I don't blame her for leaving the team. There was a lot of pressure for her. And it was very easy for everyone else to scratch their heads and not try to understand."

In fact, no one really went after Pellosma-Kelly and tried to change her mind. She had little contact with her teammates thereafter and quietly moved into the working world after graduating in 1995.

A stint as a cashier at a La Paz gas station was followed by a stream of factory jobs in the area.

"They were OK," she said. "It was fast money. But I was going nowhere."

So when her mother decided to move to North Carolina, Pellosma-Kelly figured that was as good a place as any to start over.

"It's really pretty here," she said. "It's peaceful. It's a good place where you can just think about things."

About her future with Jason. About getting a degree in environmental health. About resurrecting the dream. Her dream.

"I lost touch with Dana, but I'm so happy to hear how things have worked out," said Crone, who herself has relocated to northern New Jersey. "Everyone comes to things in their own time and on their own terms.

"If this is what she really wants to do, then we'll all be cheering for her.

"Just like we used to."

 ## Chat Room

1. Has there ever been a time when you should have listened to the advice you had been given? Explain.
2. Do your past experiences influence your future? The futures of others? How?
3. After reading this feature, do you think Hansen captured the side of Pellosma we can all relate to? What could he have done differently?

Create

1. Rewrite the story's lead or conclusion using an image or situation from a movie in which an athlete overcomes adversity, such as "The Rookie" or "Rudy."
2. Write a feature in which you interweave a key word such as *dream*. Choose a word that appeals to you deeply, as Hansen says *dream* appeals to him.
3. It is Hansen's style to write one-sentence paragraphs. Rewrite a page of this story and group closely related sentences into longer paragraphs.
4. Write an imitation of Hansen's feature using his sentence and paragraph structure but conveying your own topic and ideas.

A Moveable Moscow: A Good Hotel Near Red Square

Although Jennifer Kaye Jones has never been to Moscow, she wrote this imitation of a memoir in Ernest Hemingway's "A Moveable Feast" based on correspondence with another student who was living there. Jennifer felt she learned a great deal in imitating this model of good writing; in fact, this story played a large role in Jennifer's achievement of her Writing Proficiency designation, a graduation requirement. Jennifer imitates Hemingway's sentence structure, following the pattern of his syntax but providing her own words and meaning. The assignment required that she keep her grammar and punctuation exactly the same as the great writer's. Notice that Hemingway did not always follow the rules of grammar, especially regarding comma use.

Then there was the blustery winter. It would come in a few days before fall was even over. We would have to turn up the heat in the night against the cold and the snow would cover the domes of St. Basil's Cathedral. The walls of the Kremlin were discolored with snow and wind drove snow against the subway terminal and the Hotel Ismailovo was crowded and the windows were misted from the smoke of unfiltered cigarettes and heat. It was a crowded, decently run hotel where tourists and Muscovites were looking for a good time and I kept away from it because of the smell of cheap perfume and the sour smell of drunkenness. The men and women who frequented the Ismailovo stayed drunk all of the time, or all of the time they could afford it, mostly on vodka which they bought by the fifth. Many different viypeefki were offered, but few people enjoyed them except as a foundation to build their vodka drunks on. The drunkards were called alkashi which meant boozers.

The Hotel Ismailovo was the cesspool of Ismailovskoe shosse, the long busy highway that ran along Ismailovo Park. The parking lots of the bigger buildings, one in back of each building with two large columns, were filled with snow which was plowed by scraping and emptying into large trucks in the morning. In the morning, with the windows sometimes cracked for fresh air, we would hear them plowing and the sound was very loud. The trucks were painted black and grey colors and in the sunrise when they worked the Ismailovskoe shosse their doors and hoods looked like a Baroque painting. No one emptied the Hotel Ismailovo though, and its faded welcome sign was as cheap as its guests were drunk and despicable.

All of the sadness of the city came with the first blustery snow showers of winter, and there were no more roofs to the small dark flats as you walked by but only the blanket of white of the street and the closed doors of the small shops, bakeries, museums, and the cafes, the doctors—first class—and the dachas that the Muscovites loved to inhabit. I had a dacha at the end of the street and it was very cold and I know how much it would cost for a pile of small twigs, 1 large packet of smaller twine wrapped packages of long, pencil width pieces of pine to catch fire from the twigs and then the dried lengths of hardwood that I must buy to make a fire that would heat the dacha. So I went up to the far side of the Ulitsa to look at the roof in the snow and see if any chimneys were going, and how the smoke blew. There was lots of smoke and I thought about how the dachas might be cold and might not keep the heat and dry air in, and the fuel possibly wasted, and the rubles gone with it, and I walked on in the snow. I walked down past the Moscow River and GUM department store and the snow covered Lenin Mausoleum, and finally I came out by St. Basil's Cathedral, and walked on down Red Square until I came to a good hotel that I knew on the Ulitsa Razina.

Chat Room

1. Students go through a sort of terror when imitating "A Moveable Feast." "Is it really all right not to put a comma with a conjunction in a compound sentence?" "Is it really okay to write run-on sentences?" Why do you think Hemingway departed from the rules of grammar? Identify some of his "errors" and discuss the effect he achieves through his departure from the rules.
2. In "A Farewell to Arms," Hemingway writes about the importance of concrete words, especially the proper names of things. Identify proper names that Jennifer includes in her imitation and discuss their importance. What would be the effect of replacing them with common nouns?
3. Hemingway's "A Moveable Feast" is a memoir of the years he lived in Paris. What is a memoir? Do you think a memoir could be classified as a feature story? While the memoir tells of the writer's *personal experience*, the participatory feature tells about a writer *personally experiencing* an unusual activity or event. Discuss what makes these two actions different.

It was a pleasant hotel, warm and quiet and inviting, and I hung up my old fur lined on the coat rack to dry and put my worn and weathered lined hat on the rack in the hotel restaurant and ordered hot tea. The waiter brought it and I took my diary and started to write. I was writing about up in Sebastopol, Ukraine, and since it was a wild, cold, blowing day it was a warm day along the sea in the story. I had already seen the warm summer days come through childhood, and in one place I could write about it better than in another. That was called uprooting oneself, and it could be as necessary with some people as with other parts of growing up. But in my diary my friends and I were drinking tea and this made me thirsty again and I ordered another tea. This tasted wonderful on the frosty day and I kept writing, feeling very content, and feeling the good tea warm me all through my body and my spirit.

A young man came into the hotel restaurant and sat by himself at a bar near the entrance. He was very handsome with a face as smooth as a newly minted ruble, if they minted rubles in smooth flesh with melted snow freshened skin, and his hair was as blond and gold as a piece of straw and cut short and neat against the back of his head.

I looked at him and he bothered me and made me very excited. I wished I could chat with him at the bar, or anywhere, but he had placed himself so that he was focused on his drink and I knew he was probably having girlfriend problems. So I went on writing.

My diary was writing itself and I was having a difficult time keeping up with it. I ordered another hot tea and watched the young man whenever I looked up, or when I reached into my purse to find another colored pen when I started a new section.

I've seen you, handsome, and you belong to me now, whoever you are troubled by and if I never see you again, I thought. You belong to me and all Moscow belongs to me and I belong to this diary and this pen.

Then I went back to writing and I entered far into the events in Sebastopol and was lost in it. I was writing it now and it was not writing itself and I did not look up nor know anything about the time nor think where I was nor order any more tea. I was tired of hot tea without thinking about it. Then the section was finished and I was exhausted. I reread the section

Create

1. Locate the pages Jennifer imitated in a copy of "A Moveable Feast." Compare/contrast the two stories.
2. Differentiate between *personal experience* and *personally experiencing*, then write a paragraph of each type. Write a short memoir in which you recall a time when you lived away from home.
3. Write your own imitation of a part of "A Moveable Feast" or another work your teacher assigns. First copy the model sentences and analyze their patterns, words and meaning. Then follow these patterns but give the story your own words and meaning.

about my home in Sebastopol and then I looked up and looked for the man and he had gone. I hope he's gone and made up with his girlfriend, I thought. But I felt sort of jealous.

I shut my diary and put it inside my purse and I asked the waiter for an order of cake and a cup of Turkish coffee. After writing in my diary I was always empty and both sad and joyful, like I had recently made a visit home and then returned to Moscow, and I was sure this was a very good memory although I would not know truly how good until I read it over again in the years to come. I munched on the cake with the strong taste of chocolate and its sugary taste that the hot, black Turkish coffee washed away, leaving only the chocolate taste and the moist texture, and as I licked the frosting from my fork and washed it down with the brisk taste of coffee, I lost the lonely feeling inside and began to be optimistic and to make plans for my future in Moscow.

Work Cited

Hemingway, Ernest. *A Moveable Feast.* New York: Touchstone, 1992, 3–7.

View

The writer generally organizes the feature story with the structural elements common to the feature—a lead, a supporting narrative (body), and a conclusion; however, the writer's technique and tone create the style of the feature. Writers targeting a specific publication should study the preferred style of that publication. In all cases, though, the lead must attract a reader's attention to the feature; for example, a descriptive lead may draw the reader into a scene. Other types of leads include narrative, contrast, direct, quotation and question. Transitional words and phrases help the reader follow the flow of ideas and information through the body of the feature structure. Transitions may also set up a chronology, comparison or contrast. The writer concludes the feature in a way that will satisfy the reader's need for closure. The ending may tie up loose ends or leave a reader to ponder the future or to savor an insight.

Help

colloquialism an informal phrase common to spoken language that gives writing a casual tone.

contrast presents two different but feasibly comparable places, persons or things in the lead that associate the place, person or thing with an opposite. Contrast not only creates drama and color; it also may reveal insights.

descriptive lead serves to establish initial color or sets the stage—painting a picture of a person, place or event.

direct address uses the second person pronoun *you*, directly and conversationally to address the reader, informally bridging the writer-reader gap to create a sense of intimacy and identification.

flashback presents background information that occurred before the time of the story. A flashback can be a valuable storytelling technique, although the manipulation of time somewhat disrupts the story's flow; it should be used sparingly.

flashforward offers a part or parts of a story out of sequence to project what is expected to happen in the future and allows for dramatic scenes without the necessity of explaining background details.

impact statement creates a lead with a shocking statement or statistic that may impress or shock the reader to gain attention.

jargon refers to formal technical terminology shared and understood by a specific group of people or members of a profession but possibly not others.

localism uses a word or phrase that varies according to geographical areas.

narrative lead offers a brief anecdote, a short account of an interesting or humorous incident, to begin the feature.

nut graph usually follows the lead and conveys the story's thesis (main idea). This essential paragraph is also known as a *bridge*, *focus statement* or the spelled-out *nut paragraph*.

quotation lead arrests the reader with a well-phrased quote.

question lead generally poses a problem; the solution emerges in the feature.

rhetorical modes offer writers specific formats from which to choose depending on the desired style or effect (for example, narration, description, cause and effect analysis, process analysis, argumentation, classification and division).

slang serves to bond a group's members through informal expressions common only to that group.

transition allows the text to flow from sentence to sentence, paragraph to paragraph. These words and phrases provide verbal connective tissue.

unity serves to unify a feature through a central focus, problem or question.

Notes

1. "It's Now or Never," words and music by Aaron Schroeder and Wally Gold, recorded by Elvis Presley, was the number 10 song of the 1960–69 rock era, according to http://ntl.matrix.com.br/pfilho/oldies_list/top /lyrics/its_now_or_never.txt. This site also gives all the words to the song.
2. Jon Franklin, *Writing for Story* (New York: Plume, 1994) 151.
3. Carl H. Giles, *The Student Journalist and Feature Writing* (New York: Richard Rosens, Press, 1969) 33.
4. Franklin 151.
5. Anna Quindlen, "Weren't We All So Young Then?" *Newsweek* 31 Dec. 2001/7 Jan. 2002:112.
6. Giles 35.
7. Giles 36.
8. Dominick Dunne, "The Two Mrs. Simpsons," *Justice* (New York: Crown Publishers, 2001) 131.
9. "Statement by the President [George W. Bush] in His Address to the Nation." [http://whitehouse.gov/news/releases/ 11 Sept. 2001.]
10. Jim Meenan, "Winter's Growing Grime Prompts Cleaning Time," *The South Bend Tribune* 5 Jan. 2002:A5.
11. Bruce Garrison, *Professional Feature Writing*, 3rd ed. (Mahwah, New Jersey: Lawrence Erlbaum, 1999) 105.
12. Franklin 150.
13. Norm Goldstein, ed. *The Associated Press Stylebook and Briefing on Media Law* (Cambridge: Perseus, 2000) 22.
14. Winifred Bryant Horner, *Rhetoric in the Classical Tradition* (New York: St. Martin's, 1988) 283.
15. Horner 285–286.
16. Mark Kramer, "Breakable Rules for Literary Journalists," *Literary Journalism,* ed. Mark Kramer and Norman Sims (New York: Ballantine, 1995) 21–22.
17. Kramer 22–28.
18. Kramer 30.
19. The Irish Sports Report circulates to 15,000 paid subscribers, a national audience of serious Notre Dame fans. Hansen, designer Jack Sheard and design editor Shelby Sapusek won a Society of Professional Journalists Award in 2003 for layout/design, page 1, nondaily newspaper for The Irish Sports Report.

4

Research

Doing the research requires more time than the writing of the story.
—Carl H. Giles, "The Student Journalist and Feature Writing"

Morgan Freeman listens carefully. He interviews witnesses, watches surveillance tapes, checks computer sites and observes every possible detail from various vantage points—including a live Internet site, private homes and his parked car in the pouring rain. In the movie "Along Came a Spider," Freeman is Detective Cross, a sleuth who does his research. In fact, his detective work occupies most of the movie's playing time. Solving the crime takes just minutes at the end of the film.

Such is the case with the feature writer.

Interviewing, listening, making connections—these are all an important part of the feature writer's research process. However, before a feature writer even thinks about interviewing the appropriate people for a story, and absolutely before the writer may even consider sitting down to write the story, exhaustive background research must be completed. This chapter details research sources used by the feature writer: public records, library holdings, the Internet and personal observation.

Most writers use multiple research sources. For their story "The Sum of Two Evils," Time reporters Brian Bennett and Michael Weisskopf interviewed dozens of sources who knew Saddam Hussein's sons Uday and Qusay—"butlers, maids, business associates, bodyguards, secretaries, colleagues and friends."[1] They also "visited the sons' homes and sifted through raw material, including scores of documents, photographs, videotapes and recordings of phone taps."

Feature writers aren't pulling all-nighters. Be prepared to spend considerable time, perhaps weeks or months, in the research phase of the feature writing process.

Public Records

When staff writer Matthew S. Galbraith began work on a feature series for The South Bend Tribune about a series of neighborhood rapes, he did his detective work. According to the writer, he conducted interviews of "as many of the people involved who agreed to talk." He extensively researched public records, including trial transcripts, depositions, criminal court files (pretrial depositions, transcribed testimony and a plea agreement from several trials), and police reports. Although Galbraith specializes in investigative reporting, he combined the techniques of feature writing and investigative reporting to write "Unjustice Undone," a 16-story narrative about the so-called River Park rapes. The first story, "The Crimes," published June 2, 2002, appears in this chapter. Galbraith gains suspense and immediacy through dialogue taken verbatim from the court records, what the victim reported she said and what the rapist said.

Public documents at city halls and courthouses provide priceless details about court cases, marriages, divorces, property and family history, and may be copied for modest fees.

Library Resources

News organization libraries or "morgues," such as the one maintained by Galbraith's newspaper, generally allow public access to their holdings. He copied all news reports of incidents related to his story as well as selected items of other news that occurred the same days that seemed helpful in re-creating events. These materials helped him re-create the atmosphere, the mood, that prevailed the night of the first attack. It was "a mild spring night" and a primary election held earlier in the day "produced no surprises." In other words, everything seemed routine and uneventful until a "shadowy figure" emerged and overpowered the first victim.

News librarians create and maintain clipping files from each publication; the electronic media stores tapes of its television and radio broadcasts for future reference. News stories may also be accessed at public libraries, where print publications are kept as hard copies for a period of time; when the time period expires, the material is transferred to microfiche.

Public libraries also provide reference works on every subject imaginable. Special collections house material on subjects of interest to a particular geographical area. For example, the Hesburgh Library at the University of Notre Dame maintains a special football collection, which includes such artifacts as photographs of the famous Four Horsemen and a football carried by Knute Rockne's legendary team. Presidential libraries and private libraries maintained by corporations or special interest societies could also be useful resources for research purposes.

In 1995 a national company toured a revival of The Who's 1969 smash hit rock opera "Tommy." My editor asked me to write a *preview* that would appear prior to the opera's two-day run in our town. To be honest, I knew nothing about either The Who or their famous rock opera except that I remembered Elton John's "Pinball Wizard" as a long-ago chart topper. I knew nothing about heavy metal music except that my teenage son obsessed over what I considered to be a lot of banging and noise. Paul Stanley's face appeared handsome beneath the clown makeup, Gene Simmons' tongue occupied a "freak" category with Mick Jagger's lips, and someone named Ozzie bit the head off a bat.

I needed more knowledge for this story. Like supersleuth Cross, off to the library I went to rummage through the stacks. Whole books have been written on both The Who and "Tommy," so I kept busy for hours reading and taking notes.

Reference books abound. Feature writers nose out those appropriate for the subject at hand. Take a stroll in the reference section of any library, and you will find almanacs, bibliographies, encyclopedias, statistics indexes, "Who's Who" compilations for almost any field, city phone directories and dictionaries. As an arts and entertainment writer, I was already quite attached to the library's bound copies of The New York Times reviews. In these volumes, I found 1969 reviews of the first production of "Tommy" and also reviews of the revival.

To truly *preview* the rock opera, I rented the video of the 1975 film version and took notes as I watched. (OK, maybe this wasn't as dramatic as Cross' stakeout in the rain, but I learned a lot. I was on the trail of the elusive Tommy.) The Who's Roger Daltrey re-created his portrayal of Tommy, Elton John portrayed "The Pinball Wizard," Tina Turner portrayed the Acid Queen and Jack Nicholson appeared in his first cinematic singing role as The Specialist.

From this research, I may not have solved a crime, but I did resolve my writing problem. This is the lead I wrote for the Intermission section's cover story on Sunday, October 8, 1995:

> Groomed by their manager for the Mod scene in the 1960s, they wore tight clothes with pop-art designs. Their propensity to smash up their instruments at the end of each concert put them a half million dollars in debt. "You went to see The Who and didn't know what was going to happen," Elton John has said of the group.

My research revealed that early critics viewed the rock opera's "child molestation, sadism and LSD-sex therapy as perverted. . . . Yet 'Tommy' rose to No. 4 on Billboard and went gold." Its initial U.S. tour "included a stint at Woodstock, which brought 'Tommy' back this time to No. 1." The 1993 volume of The New

York Times reviews gives the contemporary critical view: "Frank Rich couldn't praise The Who's 'Tommy' enough."

I read the script for the play (really just song lyrics) so that I could convey, in my own words, the rock opera's plot to my readers. I interwove specific facts gleaned from research with critical opinion. Some opinions surfaced frequently enough to be considered common knowledge: "The Who's sound and stage antics brought the first blasts of heavy metal." Notice that direct quotes and paraphrase require attribution, while information that has become general knowledge does not.

The Internet

Is the Internet your first stop when you're assigned a research project? While the infinity of online resources may seem more accessible and diverse than tangible library holdings, it's a mistake to circumvent the library jaunt. Print sources generally offer fuller, more reliable information. Never assume everything in print, in hard copy or online, to be the gospel truth, of course. That all printed material is necessarily true and accurate is a common misconception. For this reason, magazines and newspapers have long held to a rule of three for controversial stories: A fact must be verified by three sources before it goes to press. Checking sources occurs daily in the mainstream media; the bad name attached to the press generally originates with tabloid reporting. For this reason, I'd never extract material from a tabloid report and quote or paraphrase it in my own story.

Having said this, the Internet remains a rich and convenient research tool. Search engines such as Google and Yahoo allow users to broaden, narrow and define searches of infinite amounts of material. If you've done much research on the Internet, you know that you need to be careful about the sites you access. When I was researching a story on England's current Royal Family, I found well-researched facts on British media sites and some outrageous material on sites created by private citizens. One site, which turned out to be pornographic, morphed a photo of Prince Charles in a degrading way.

Some libraries have purchased and installed software with such periodical indexes as INSPIRE. These programs allow for narrowly defined searches on specific topics. Online reference works such as dictionaries and encyclopedias can also speed up the research process. Ask what's available while you're in the library or check the library's home page.

Personal Observation

You are not writing fiction; research yields the factual backbone required for all media stories. Clearly, research also provides the detail that will bring the feature story alive to the reader. The writer becomes the eyes and ears for the reader. You are a reporter, and what you observe should be recorded in your notes so that later, when you begin to write, you can describe a scene, an atmosphere, a person in such a way that the reader can share in the experience.

"The reader couldn't be there, but the writer was," writes Abraham Aamidor in "Real Feature Writing."[2] "He or she has to show what it (whatever it was) was like, in graphic detail."

This show-and-tell requires excellent skills of observation, the meticulous recording of details in your notes and the use of these notes to report and describe, to make the reader "see and hear and feel" as you write. Whereas preparation to write a participatory feature might require *direct observation*, taking part in the action, most reporting involves *indirect observation*, surveying the scene and the action and listening to the actors. The Colm Feore story in Chapter 3 and this chapter's Sean Savage story combine direct and indirect observation (the writer observed Savage but later they shared soft drinks and social conversation).

Observation engages all the writer's sensory perceptions—touch, taste, sight, sound and smell. If you're at a racetrack, can you smell the burning rubber from the car tires? If you're writing about chocolate on Valentine's Day, taste the stuff. Touch the fabric of a sofa, describe its color. Look for the typical as well as the unusual. Listen to what people say and write down their words for quotes, but listen to their voices, too. Notice what people wear.

Later, when the research phase of the process has been completed and the writing begins, the writer—like the detective—considers and interprets the gathered "evidence." For now, consider the aspects of persons, places and events you should observe:

Voices

The meaning of a person's words may be conveyed through the pitch, quality and volume of the voice as well as the pace of speech. Look back at Molly Donnellon's "Making Stone Soup" feature in Chapter 1. John speaks to her "quietly. His speech is slow and sounds as if he put forth great effort just to get the words out." She describes Mary's voice as proud and loud. The reader can infer individual personalities from these details.

Body Language

Some of us are just naturally people watchers. We don't need to be told that body language reveals as much about individuals as the words coming out of their mouths. Observe gestures, facial expression, body language (such as arm and leg crossing, leaning toward or away from someone or something), movement (how someone walks, sits, stands), "the gaze" (where and how someone is looking—directly at you, with eyes drifting upward or staring downward).

Andrea Dominello reports detail about body language in the lead for her "Bomber" feature in Chapter 1. She describes Jack Keller, at age 17, walking into the kitchen to see "his parents crouched over the radio." Later, she observes a picture of Jack's flight crew, with Jack "standing tall."

Clothing, Hair Style and Props

Notice what people wear and how they wear it. The Troy Donahue story in Chapter 2 mentions the actor's signature red sweater; in a feature in Chapter 7, Susanne Alexander describes 58-year-old Dale Huston as "getting gray around the edges," a reference to his graying hair. In Chapter 3 the actor Colm Feore arrives for an interview in a T-shirt and jeans. In Chapter 5 the feature about Tom Wopat describes the "crisp coral shirt and white shorts" he wears for an interview at a lakeside restaurant. Color, fabric, detail, detail!

Props, or the paraphenalia a person chooses to hold or to display in personal environments, add color and interest to feature writing. In her feature about Sean Savage in this chapter, Holly M. James observes a picture of an orangutan wearing a cap and gown on a college professor's office door; inside the office, she observes piles of ungraded papers and a stale chocolate chip cookie.

Proxemics

In 1959, anthropologist Edward Hall "popularized the study of spatial perceptions and use of space—calling it proxemics—in his classic book, 'The Silent Language.' "[3] Spatial signs, signals and cues to observe include distances between people (body space), crowded spaces, organized space such as that in a library or parking lot and the size, shape and utilization of office and home space.

Hall categorized human spacing behavior (in the theater it is called "blocking") into four bodily distances: *"intimate* (0–18 inches), *personal-casual* (1.5–4 feet), *social* (4–10 feet) and *public* (10 feet and beyond)."[4] Whether individuals appear to be excluded or included in a group should also be noted.

Journalists and detectives aren't the only professionals who must develop observational skills. A dermatologist at the Yale Medical School who was the director of a program for first-year medical students was unhappy with his students' ability to make diagnoses. After all, serious repercussions might result from a misdiagnosis, a life-or-death matter for patients. With the help of the curator of education at the Yale Center for British Art, he instituted a three-part program that has become part of the first-year syllabus for all medical students at Yale.

In this program, students view paintings in groups, record their observations, then present their paintings to members of the other groups. When someone "catches" a person interpreting instead of describing an observed detail, the offender is called out for the infraction. This exercise helps students understand the nuances of difference among observation, description, reporting and interpreting.

Journalists, like detectives and medical professionals, observe, record and later interpret what they see, hear and feel around them. Interpretation will be explored in Chapter 6.

The reporter's curiosity also motivates successful historians who base their careers on in-depth research. In this chapter's Click Here, presidential scholar Max J. Skidmore cautions would-be professional writers about the current climate regarding attribution and plagiarism. As you research your features, remember the rules of fair play in terms of borrowing others' words and ideas.

Click Here: Those Useful Citations

Max J. Skidmore, University of Missouri Curators' Professor of Political Science at the University of Missouri–Kansas City, contributed the following words of advice about plagiarism.

Max J. Skidmore
Author of "Those Useful Citations"

Plagiarism is a cardinal academic sin.

To present work from others as one's own violates a principal rule of scholarship. Simply put, it is dishonest. Don't be guilty of it. Anyone who uses someone else's words is required to enclose those words in quotation marks and to give the source. Similarly, any writer who uses ideas from someone else is obligated to give credit to the other person.

This is not to say, of course, that one must attribute every idea and every phrase. If one were to write that something were necessary in order that "government of the people, by the people, and for the people shall not perish from the earth," it would be a good idea to give credit to Lincoln, who used the phrase in his "Gettysburg Address."

But some things are matters of general understanding. It is hardly necessary to provide a reference every time one writes "government of the people," which has entered the language and become common usage. That the Declaration of Independence bears the date of July 4, 1776, or that Thomas Jefferson was its principal author, need not be cited. It would be foolish and pedantic to document what everyone knows. On the other hand, as the late humorist Will Rogers allegedly said, "It ain't what people don't know that's so dangerous—it's what everyone knows that just ain't so!"

Attribution of words is not the only concern; one may also need to attribute ideas. There is a difference between saying "the grass is green," which everyone really does know, and saying, for example, "Social Security is going broke." The latter statement requires documentation, not to avoid plagiarism, but to be meaningful. Many consider it a factual statement about Social Security, but only because they have so frequently heard frightening allegations.[5]

The popular historians Stephen Ambrose and Doris Kearns Goodwin received criticism for failing to give adequate documentation to other authors for some passages. They admitted that, in some instances, they should have taken care to identify passages more fully. At best, they were guilty of considerable negligence; in any case, their reputations suffered greatly. Nevertheless, they did cite their sources and their works clearly are not "plagiarized." Anyone who delves deeply is likely to discover similarities in a number of works on a given subject. Some inconsequential, unintentional similarities in passages never raise an eyebrow.

No doubt this may come as a surprise, but there is considerable pettiness in the world of academe. It is fairly common to hear sneers directed toward "celebrity historians" who win the "obligatory Pulitzer Prize." Plagiarism, legitimately a cardinal sin, has a less savory twin: another (albeit unofficial) cardinal sin, that of popularity, or prominence.

Some years ago, Stephen B. Oates, a distinguished historian and Lincoln specialist, was the subject of a surprise attack. As he put it, "A literary critic, a professor of classics, an associate professor of criminology, and two historians" argued that he had plagiarized his Lincoln biography, "With Malice Toward None."[6] Oates defended himself vigorously and issued a thoughtful rebuttal in which he discussed "confusion about what constitutes plagiarism." It deals with the theft of material from others. The criticisms in his case—as is frequently the case when prominent figures are the targets—dealt with "miniscule fragments to the exclusion of everything else." His challengers, he said, ignored how different his book was from the others.[7]

Is plagiarism a matter of concern only to nitpicking professors?

Hardly, as a glance at some random news items will demonstrate. A few years ago an up-and-coming young writer, Stephen Glass, found himself in trouble and without a job when several magazines discovered that he had made up all or parts of articles that they had accepted on good faith and published. Glass also was held up to public scorn when headlines across the country revealed that Forbes, George, Harper's, Rolling Stone and The New Republic magazines wanted nothing to do with him any longer, and that The New York Times rejected one of his articles that it had planned to publish.

In Indiana, both the Indianapolis Star and the Indianapolis News discovered that their television columnist had lifted material from another newspaper. After apologizing for what he minimized as a stupid mistake, the columnist, Steve Hall, received a suspension without pay. When he returned to work, it was in a lesser position. At least he still had a job. Many of those who engage in plagiarism are not so lucky. The New York Times reporter Jayson Blair "took a brief leave of absence" in April 2002 when it was discovered that he fabricated quotes and stole material for his cutting-edge stories.[8] Following the leave, Blair was again accused of fabricating and stealing stories, forcing his resignation. The New Republic also fired an associate editor, Ruth Shalit, because of allegations of plagiarism. Mike Barnicle, longtime Boston Globe columnist, became a former Boston Globe columnist when someone pointed out that he had included in his column, as his own work, material from a book by George Carlin.

For students as well as journalists, a clear lesson emerges. The consequences of plagiarism can be major. Understand what requires documentation. Don't take chances. When in doubt, provide attribution!

"Injustice Undone: Part 1: The Crimes"

Matthew S. Galbraith, a South Bend Tribune staff writer, describes the process of writing his series: "Another reporter suggested we serialize the story of Richard Alexander. It was a perfect fit in terms of a main character who overcame adversity and a storyline about the universal injustice of an innocent man being imprisoned. The *narrative style*, on the rebound after widespread use in the 1960s and 1970s, tells the story as it happens, through the eyes of those involved, flavoring the facts with observations to lend depth, feeling, emotion and meaning to the characters' actions.

"In this case, a certain amount of *investigative reporting* was necessary. Investigative reporting probes accusations of wrongdoing, especially by public officials and prominent figures, and seeks to assign responsibility with strong factual support to prove the allegations. Narrative journalism was a new writing style to me. Our newspaper had done this type of project a few times, but I had not been involved. My specialty is investigative reporting; I had worked on several of these projects either alone or as part of a team.

"The result was a mix of styles, with the narrative complemented by the investigative in most of the stories and the other way around in a couple of stories, depending on the content. The research, I discovered, was much the same. The challenge was to differentiate which information was most useful to the narrative aspects and which to the investigative aspects. Writing was more difficult. My investigative instincts told me to link related information to form conclusions, which frequently meant jumping ahead in the chronology. My editor sent me back to the writing process to trim the stories. Of course, I protested. I feared readers would lose track of the ongoing events without some guideposts to help them understand the main points and then lose interest. In the end, I obeyed the editor and saw that the writing plan could work with some timely references to earlier, significant events as the stories unfolded."

These words appeared in reverse print on a black box when the story was published in The South Bend Tribune on June 2, 2002: *The true story of the River Park rapes. This unusual 16-part series examines terror, justice and one man's redemption. These stories, by the very nature of the crimes involved, concern serious adult subjects. Parents are cautioned.*

In 2003 the story won a first place award from the Society of Professional Journalists and a third place award in the Indiana Associated Press Managing Editors' Newswriting and Photojournalism Contest.

Nothing seemed out of place on the mild spring night. Earlier, voters had gone to the polls in a primary election that produced no surprises. A nearby overhead light provided an illusion of normalcy.

House-sitting in the 500 block of 28th Street for a friend who was out of town, the 37-year-old woman went about the same tasks as she might at her own home after arriving about 9 p.m. The date was May 7, 1996.

She straightened up the inside of her car and washed the windshield. Then she grabbed an armload of books and clothes to take inside for the overnight stay.

She entered the front door and went to the kitchen, where she set down her purse and belongings and let the dog out the back door. It was because the dog needed watching after a surgical procedure that she was there. She hung her jacket on a large brass hook in the living room.

Unseen in the closet, a shadowy figure was ready to pounce on the unsuspecting victim. She would gamely put up a fight before being overpowered by the intruder's strength, agility and illicit intent. Had he been inside the house when she arrived?

"I want your money, b———," he demanded in a slurred, streetwise slang.

She backed up to a couch; he advanced toward her. She dropped to the couch and kicked him between the legs. When he put his hand down to protect himself, she noticed that he was carrying a knife.

Though he wore a gray ski mask, she could tell he was black, about 5 feet 10 inches and 170 pounds, probably in his 20s.

Trapped, she made a run for the back door. He blocked her, moving so deftly as to make any chance of escape a physical impossibility.

"No, you're not going anywhere," he said, then he pulled her into a bedroom off the kitchen and ordered her to sit on the edge of a bed facing the kitchen.

After blindfolding her with a scarf and removing his ski mask, he fired several questions at her. "Where's your money?" "Where's your purse?" "Is there any other money in the house?" "Are there any drugs in the house?" "Is there a gun in the house?"

When the scarf flipped from her eyes, she observed him for several seconds silhouetted against a light in the kitchen. It was not necessarily a good look because he had removed her glasses, but he noticed.

"Well, you realize I'll have to kill you now because you've seen me," he remarked.

After searching her purse, he raped her. Humiliated, she wanted to get it over with as quickly as possible. When he put his pants on and looked through her purse a second time, she decided to make him think she was unbalanced. She wanted him gone. She wanted the ordeal to end.

"Well, what are you waiting for?" she asked sarcastically. "Wait, wait, wait. Just like a damned man. I have to wait for everything."

The ploy seemed to work because he left, but she omitted the sexual assault when she first talked to police, describing only a home invasion. The damage was done, and she didn't want to upset her mother. She figured the police could do nothing about what had happened to her.

A Second Attack

May's last week in 1996 felt more like April with chilly, 50-degree nights and rain showers that drenched most of the area's Memorial Day parades and continued throughout the next day, May 28.

A little north of the earlier home invasion and sexual assault in River Park, a 25-year-old woman was putting her 6-year-old nephew in her pickup truck after picking him up from a baby sitter who lived in Courtyard Place apartments. It was about 10 p.m. and raining. The youngster was sleeping.

A man approached on the sidewalk on 26th Street. He was stalking her, perhaps sizing up her smallish, 5-feet-1 frame and added vulnerability of carrying the child. She saw him closing and knew he was coming for her, as if he had hunted her. She averted her eyes briefly from him and secured her nephew into the passenger side of her Ford Ranger.

The man came from behind, grabbed her around the neck and demanded money. He had a gun. After taking about $10 from her purse, he took her behind a nearby house with a fence in back. Standing there in the ankle-deep wet grass, her nephew asleep in the truck, the victim was as powerless as the child.

"Wait, where's my keys?" she asked frantically at one point, worried about her truck and the child inside it.

"I don't have your f———keys," he growled.

There was some lighting behind the house, but she could not tell whether it came from a streetlight or a nearby porch light. "I could see the rain hitting the grass. I could see that glittery effect," she said later.

Her attacker was a black man, 20 to 25 years old, about 5 feet 11 inches, 160 pounds. She thought he wore gloves. Though she didn't get a good look at his face, she remembered his baggy pants and the way he spoke, the slang he used, especially when he muttered the word "b———."

Her passivity turned to active resistance, however, when he pushed her down to her knees next to the fence and tried to force her to perform oral sex on him. She resisted; he hit her; she bit him.

"Don't bite me, b———," he said three or four times. The woman began to scream, and they took off in opposite directions.

Sylvia Agnone, who lived in the house in the 400 block of South 26th Street, watched these events from her bedroom window while her husband called the police. She was about 15 feet away from where the man—who by then had forced the woman to her knees and unbuttoned his pants—was hitting her.

Afterward, the attacker passed by her window, only about 3 to 5 feet away from where she stood inside, and she saw his face. She also described him as about 5 feet 9 inches and 140 pounds.

Coming Monday: The Reaction

 ## Chat Room

1. Galbraith calls this story "a mix of styles." Identify narrative (feature) elements, then identify elements more typical of investigative news reporting.
2. The writer believes that the narrative *complements* the investigative and that the investigative complements the narrative. Find examples of both.
3. Galbraith hoped "to differentiate which information was most useful to the narrative aspects and which to the investigative aspects." How did he do this? Point out specific information.
4. The editor wouldn't allow Galbraith to "jump around in time" even though readers might need "guideposts" to understand the story and lose interest in a chronological presentation. Identify places where the writer refers to "earlier, significant events." Do you feel the story's organization aids your understanding? How does its organization affect your interest?

Create

1. Review a detective movie. Detail the ways the detective gathers information and clues.
2. Galbraith's series tells the story of the conviction of an innocent man for the River Park rapes. A half-hour segment of "Forensic Files" on Court TV told the same story. Television programs use different techniques than newspapers to tell stories. Watch the segment if you can.
3. Make separate lists of the storytelling techniques used by television and newspaper.
4. Make a list of sources evident in Galbraith's story.

FEATURE STORY

"Roxie Hart"

As a research assignment, I divided my class into groups, each responsible for a different aspect of the movie "Chicago": movie reviews, the movie cast, the stage play. After the groups presented their findings and the others took notes, we took a field trip to the local movie theater to view the film. With an incurable reporter's curiosity, I began to do research of my own. From that research I wrote the following *color feature* (a secondary story that backgrounds a primary news or feature story). This story might background a news story about the Academy Awards presentation.

"Nobody walks out on me," says Roxie Hart as she plugs boyfriend Fred Casely. Despite his protest, "Sweetheart—" she sneers, "Oh, don't 'sweetheart' me," as five shots ring out.

In John Kander and Fred Ebb's 1975 stage play "Chicago" and its 1997 revival, the merry murderess evokes little sympathy. Not so in director/choreographer Rob Marshall's film, which received the Academy Award for Best Picture on March 23.

Although Renee Zellweger's Roxie objects to a press conference suggestion that she regrets "doing" Fred ("Are you kidding?"), her waifish figure and pound puppy eyes make her unique in a long line of real and fictional "jazz killers," the stage and screen Roxie Harts who now span a century.

Zellweger has become Roxie for the 2000s.

Although the Oscar eluded her, Zellweger took home a Golden Globe for best actress in a musical earlier this year.

The saga of Roxie Hart began in 1924 when Chicago Tribune reporter Maurine Dallas Watkins covered the murder trial of Beulah Annan, a garage mechanic's wife who found fame after she shot her boyfriend and left him to die while she played "Hula Lou" on her Victrola. Beulah's coldness emerged as she tried initially to blame her husband for the murder, then pushed the Cook County trial forward with a fabricated pregnancy.

Watkins quit her newspaper job and enrolled at Yale, where she wrote the play "Chicago" for a writing class. Beulah became Roxie Hart in Watkins' 1926 Broadway hit, made into a 1927 silent film supervised by Cecil B. DeMille and starring bow-lipped, big-eyed flapper Phyllis Haver. A studio still of Haver with tousled blonde curls foreshadows Zellweger's Roxie to come.

However, other Roxie Harts waited between.

The 20th Century Fox film "Roxie Hart" became a star vehicle for Ginger Rogers in 1942. She plays Roxie as a gum-snapping, head-butting showgirl in an apparently deliberate departure from the elegant, evening-gowned image she established partnered with Fred Astaire. Rogers' dance talent is relegated to one jailhouse Charleston routine and a tap atop the cellblock stairs.

Those who think the current movie is corny should take a look at the video of the 1942 version, released in a classic film series in 1995. All the usual suspects are there—a sleazy Billy Flynn (Adolphe Menjou) defends Roxie at the infamous press conference and in the prosecution-baiting courtroom scene; Matron Morton breaks up a cat fight (back-dropped by screeching sound effects) between Roxie and a Velma Kelly prototype by knocking their heads together; husband Amos Hart does a sad tap dance outside the cellblock after he is denied entrance for the big dance scene.

You can see it comin'.

1960s stage and television producers itched to adapt Watkins' script, but the aging screenwriter had qualms about unleashing Roxie again. Watkins' coverage of the Annan trial hints at her attitude toward the role celebrity played in letting someone get away with murder. For a story about jury selection for the trial, Watkins' headline read, "Wanted: Twelve Good Morons."

Watkins' satiric critique of the American media and judicial system got lost in the 1942 film. Rogers' callous chorus girl gets off, and Amos is arrested for the murder, perhaps a concession to the old studio code's dictate that someone had to be punished. Roxie ends up happily remarried with a car full of kids in the frame the movie added to Watkins' story.

Producer/choreographer Bob Fosse had to wait until Watkins' death in 1969 to get the rights to co-author the musical adaptation with Kander and Ebb.

Fosse's ex-wife Gwen Verdon brought Roxie back to Broadway. A flashy redhead best known for her shockingly sexual Lola in "Damn Yankees," Verdon ended her Broadway career wearing a bobbed blonde wig and baring her midriff for the "Hot Honey Rag" finale with Chita Rivera (Velma Kelly).

Fosse protege Ann Reinking replaced Verdon in the long-running show. The statuesque Reinking, leggy and limber, wore a little more (white gloves, top hat) while wearing less (a garter, thong and low-cut vest). Her seemingly effortless 180-degree extensions, precise moves and sexual struts epitomized the Fosse dancer. She snapped her heels with class; Haver and Rogers had only snapped their chewing gum. Reinking came back as Roxie in the 1990s "Chicago" revival, but the musical remained an ensemble act, with Tony Awards given to both Reinking and Bebe Neuwirth (Velma).

Its Screen Actors Guild Award for best ensemble acting attests to the fact that the current film keeps the integrity of its stage predecessor. It is not a star vehicle for Zellweger—or anyone else in the cast. Nevertheless, like Rogers before her, Zellweger works against type in taking the role. Zellweger made her name as the loyal, mousy wife in "Jerry Maguire" (1996), the victimized title character in "Nurse Betty" (2000) and the wronged-again but more weighty star of "Bridget Jones's Diary" (2001).

Bill Condon's screenplay for the movie gives Zellweger's Roxie initial innocence, but in this film she does not end up with a reformed Tom Cruise or a compassionate Colin Firth. An extreme closeup of Zellweger's eyes opens the film, then cuts to a neon image of Chicago glitz. This is the direction the film takes: Innocence will be consumed in the flames of institutional corruption.

Whereas previous Roxies gun the boyfriend down when he threatens to walk out, not a compelling reason to shoot someone, some may feel Fred Casely (Dominic West) gets what he deserves. For the first time, the audience sees his propensity for violence—after some pretty rough sex, he treats Roxie rudely, shouts in her face and shoves her around. He also reveals that he flat-out lied to her to "get a little of that" and threatens her. "Touch me again, and I'll put your lights out," he promises just before he throws her into a wall.

Zellweger's Roxie is not the cold-hearted Beulah Annan, who watched her victim die while she put some jazz music on the phonograph. Instead, Zellweger's petite little Roxie shares her dream of a tasteful act with the big lug who, in response, crumples her against a wall.

Rogers' Roxie never shed a tear; Zellweger gushes them. Forget Reinking's thong. Zellweger coos "Funny Honey" as she and her vintage peach satin costume melt languidly onto the piano top. At the end of the number, when her husband betrays her, Zellweger pounds the piano and grabs at her skirt in a toddler's tantrum.

To further prod viewer emotions, John Myhre and Gordon Sim (who won Oscars for art direction) show a grimly authentic prison, dark and cold. Zellweger's Roxie inquires whether there's "something wrong with the heat." She asks Matron Mama Morton (Queen Latifah) for a "couple extra blankets" that she thinks Mama may have "tucked away." We must marvel at her naivete. This is a corrupt prison system, and we just saw Mama Morton pull bribe money from inmates' garters.

Each version of Watkins' story added murderesses to the cast, but only Marshall's film puts one of them on the gallows. The Hunyak (Ekaterina Chtchelkanova), the only prisoner in "Cell Block Tango" to declare herself "Not guilty," loses her last appeal. Zellweger watches her hanging, then confides to Billy Flynn (Richard Gere), in her little girl voice, "I'm scared."

Finally, though, as Flynn says, "It's all a circus . . . a three-ring circus . . . the whole world." Audience sympathy is stretched when Roxie, who moments earlier testified that she killed Casely to protect the life of her husband's unborn baby, faces Amos in an empty courtroom. She finds it incredulous that he believed her story. "What do you take me for?" she asks him. "There ain't no baby." Of course, the savvy knew all along that hers was "an act with lots of flash in it" but a lot of truth in it, too.

The stage revival of "Chicago" came on the heels of the Simpson and Menendez trials, its popularity said to reflect American cynicism about the media and the law. Zellweger's Roxie finally portrays innocence lost.

In the last number, Zellweger comes out armed with a machine gun for "Nowadays/Hot Honey Rag." Her corruption complete, Roxie still has something Americans admire—the triumph of Zellweger's 21st century Roxie is resiliency. A New York Times headline on April 6 reflected what America craves nowadays. It read, "Some Resilience in a Shaky World."

FEATURE STORY

"Sean Savage"

As a student in my feature writing class, Holly M. James chose to write about one of her professors for the personality profile assignment. She made arrangements to interview him in his office so that she could soak up the atmosphere of his private space. Her observation of the quirky, the unusual, make this feature fun to read. Holly offers details as braille for the blind, so that we can all be in her professor's office and see what she saw.

Papers lay helplessly, begging to be graded, or at least to be stacked somewhere other than on the floor. Piles of books, like miniature mountains, line the little office. On the office door, an orangutan wearing a cap and gown asks inquisitively, "What? You haven't been to Harvard?" To join the primate, other items are stuffed haphazardly in front of a window. On the desk sits a stale chocolate chip cookie.

Keeping watch over this little world of books, papers and primates is Sean Savage, author of "Roosevelt: The Party Leader, 1932–1945" (1991) and "Truman and the Democratic Party" (1997). Savage fans may sign up to receive an e-mail from the Amazon.com's letter-beaked bird when the next "sean savage" release arrives. An associate professor of political science at Saint Mary's College, Notre Dame, he is currently completing his third book, "JFK, LBJ, and the Democratic Party."

Savage grew up in Worcester, Mass., situated between Boston and Springfield. He attended Assumption College in his hometown, double majored in political science and history, and minored in English. He finished his undergraduate studies in three years; in addition, he was named valedictorian of his graduating class. He was moved toward journalism at first, but finding a permanent position in that field was difficult. He then became interested in law, but money was an issue. "I couldn't afford it," he said during an interview in his littered office. "So I just followed the money trail." Grants and scholarships led him to Syracuse University, where he earned a master's degree, and Boston College, where he received a fellowship named after Tip O'Neill and earned a doctorate in American politics.

He swayed in the direction of teaching.

Throughout his schooling, Savage was involved in student government and in various political campaigns. "Because of me, Reagan carried Massachusetts in '84," he says. Savage has also worked in the campaigns of Walter Mondale and Michael Dukakis and interned with former congressman Joe Early. "I don't know how much I benefitted Mondale by working with his campaign, though."

With the credentials Savage carries, it is possible to imagine him as stuffy. But stuffy he is not. "If I wasn't teaching, I think I'd be a monk," he said, because "they basically pray, eat and hang out in a big, brown robe. Not a bad deal." He reached into his little refrigerator and pulled out two Diet Crush cans. "Hey, what has 400 legs and 11 teeth? The front row at a Willie Nelson concert." He passed the cans over a desk covered with papers, Post-it notes, and a picture of two cats. "Those are my cats, Bess and Harry Truman. Bess left us, so Harry is now a bachelor. Oh, well. Now he can leave the toilet seat up permanently."

Savage holds a special connection to his cat Harry Truman. "When I first arrived here in South Bend, I needed inspiration. I saw the cat all the time, so I named him Harry Truman, after my favorite president. He was a constant reminder for me that I had something to focus on."

Savage holds an even more special connection to Harry Truman, the man. "There have been three miracles in history: the parting of the Red Sea, the New York Mets in 1969, and Harry Truman winning the presidential election in 1948. What an upset." Savage said he admires the underdog victory of Truman as well as the policies that Truman championed.

Teaching presidential politics is Savage's forte. He has published two books and several articles on the American presidency. He was first smitten with the American presidency because of John F. Kennedy—not just Kennedy's Massachusetts residency but more his leadership style. Savage noted that, because Kennedy was the first president to hold regular televised press conferences, many felt a real connection to him. He recounted that Helen Thomas, one of the few women allowed in the White House press room, asked JFK, "What are you doing for women in America today?" JFK responded, "I'm doing *all* I can." Savage was amused for four minutes straight.

Savage loves to intertwine humor and politics. He explained that from early times scandals involving major political figures have involved sexual relationships. He laughed at former President Bill Clinton for having too many Little Debbie's, "both of the snack and human variety." Savage may love politics, but not the idea of political correctness.

Starting with Clinton, Savage went on to unveil many untold secrets of past presidents. He likes to study the presidency, especially "what goes on behind the scenes." Savage told an amusing story involving J. Edgar Hoover's investigation into the song "Louie, Louie." Hoover "didn't know what the song was about, so he ordered a federal investigation." When asked about the outcome of the investigation, Savage shrugged and said, "We still don't know." He believes that former vice-president Al Gore's beard makes him "look like an unemployed psychologist." This left him chuckling.

He stated that Bush is smart enough to understand his own limitations. Savage pointed out that Bush has surrounded himself with specialists "to cover the areas he can't cover himself." Savage pointed out one of Bush's flaws. "He's a little too chummy with the oil and gas market." Then he added with a laugh, "You know what Texans say—propane is God's gas." He leaned over his desk, laughing at his own joke.

"I like teaching politics," he stated when he recovered. Politics is "unpredictable. There is always more to learn." He referred to the 2000 presidential election with a laugh: "The losing candidate ended up winning the presidency."

Savage clearly loves the world of politics, and teaching is the exercise of his love. "It's probably a geeky thing to love," he said. Then, with a big smile, he added, "But you know what the Bible says. The geek shall inherit the earth."

We drank our Diet Crush.

Chat Room

1. To write the "Roxie Hart" story, I watched the movie "Chicago" four times. Identify other films I watched. What other sources employed in this research project can you identify?

2. Color features tend to be brief. Does the "Roxie Hart" feature have elements of other feature types? Locate other "hybrid" features in newspapers and magazines. As you look, ask yourself to what extent published features take hybrid types. Discuss this in your peer group.

3. Holly's description adds humor to her feature. Highlight places in her story that allow you, the reader, to be there with her. Why do some of these observations achieve a humorous effect?

4. What could Holly add to the "Sean Savage" feature to further put the reader in the story?

Create

1. Choose a band, other than The Who, that has made a movie (i.e., The Beatles). Go to the home page for your school's library. Using the search engine, print out an article about the group or its music style from a scholarly journal. Next, look for information about the group on a search engine such as Google. Go to a couple of the sites that come up and print out the information. Finally, go to http://encyclopedia.com and search for your group.
2. Take notes as you watch a movie made by your group. Write a feature based on your research, including that for exercise 1.
3. Sit in on a trial at your local courthouse. Observe and take notes. Obtain a transcript of the trial. Write a feature based on your research.
4. Boy bands seemed to be the craze until "American Idol" showed up in everyone's living rooms. Choose the latest craze and find five or six Internet stories about the person or the group. How many of these sources are reliable? How can you tell? Who wrote the stories?

View

Research is detective work. You have to use all five senses in order to be a good detective—sight, sound, smell, touch, taste. Uncover as much information as possible—public records are extremely reliable. They offer obvious facts as well as clues to other sources—for instance, names of acquaintances and addresses. The media may be a viable source, but avoid tabloid reporting. Verify at least three sources if you're dealing with controversy. The Internet may provide dependable information, but be sure the Web site is legitimate and the information is accurate. Also, feature writers make personal observations not only to contribute information but also to add interest through giving a personal touch to the feature.

📖 Help

attribution   the use of quotation marks and source citations to acknowledge words borrowed from someone else; ideas not part of general knowledge also need to be acknowledged.

direct observation   in reporting a feature story, writers observe while taking part in the action.

indirect observation   in reporting a feature story, writers survey the scene and the action while remaining uninvolved.

plagiarism   the dishonest use of another person's words or ideas, a violation of a principal rule of scholarship and reporting.

preview   a story that provides readers useful information about an upcoming event.

private library   a resource center such as a presidential library or a library maintained by a corporation or special-interest society.

proxemics   the study of spatial perceptions and use of personal space, derived from anthropology, that enables observant writers to gain insights about others.

public records   documents available to reporters and the public at large, including trial transcripts, depositions, criminal court files and police reports.

search engine   a commercial server, such as Google or Yahoo!, that allows users to broaden, narrow and define searches of infinite amounts of material.

Notes

1. Brian Bennett and Michael Weisskopf, "The Sum of Two Evils," *Time* 2 June 2003:37.
2. Abraham Aamidor, *Real Feature Writing* (Mahwah: Lawrence Erlbaum, 1999) 129.
3. "Proxemics," in David B. Givens, *The Nonverbal Dictionary of Gestures, Signs and Body Language Cues* (Spokane: Center for Non-verbal Studies Press, 2002), <http://members.aol.com/nonverbal2/diction1.htm>. 8 Sept. 2002.
4. Givens. 8 Sept. 2002.
5. See my book, *Social Security and Its Enemies* (Boulder: Westview Press, 1999), for a counterinterpretation.
6. Stephen B. Oates, *With Malice Toward None* (New York: Harper Perennial, 1994) xvi.
7. Oates xviii.
8. Nancy Gibbs, "Reading Between the Lies," *Time* 19 May 2003:57.

5

Interviews

I like to listen. I have learned a great deal from listening carefully. Most people never listen.
—*Ernest Hemingway*

Tom Wopat walked off the stage after a rousing curtain call. He bent over the drinking fountain by the stage door as I arrived for a hastily arranged interview. But I was quicker than the associate producer who had made the arrangement. He didn't know he was to be interviewed. A look of "What the . . . ?" passed over his face. Then he got his drink.

With the same sweeping motion he had used moments before as he sang, "When the wind comes sweeping 'cross the plain," he invited me to have a seat on a backstage sofa.

This happened in 1983. It was the first of a half dozen interviews Wopat gave me over the years, and I am fortunate that I got any interviews at all. Impromptu interviews to which the interviewee has not agreed usually don't work out very well. Luckily for me, Wopat was a real trouper. He was sweaty, tired and hankering to go off to an opening night cast party. I was barely prepared. I had come intending just to review "Oklahoma" but couldn't pass up the offer to interview him.

I always prepare for interviews, and I had not gone through the steps I usually do before the unexpected Wopat interview. This chapter details the art and science of interviewing.

Types of Interviews

Only a few will ever have the opportunity to chat over lunch with Tom Wopat or share dinner with a dieting Al Pacino; when the feature writer is allowed at the table, so to speak, the writer shares that moment with readers. The purpose of all interviews is to provide the writer with good quotes and observations to make the reader's vicarious experience as full and sensory as possible. Quotes and observations gathered during interviews, added to the preliminary research, should produce insightful stories that may, at best, in some way enhance the reader's understanding of the meaning of the larger issues in life, such as success and fame.

Not all interviews lead to profile stories, of course. Interviews add color and information to other kinds of features. For example, Kaitlin E. Duda's trend feature in Chapter 10 gains human interest from its inclusion of expert commentary on reality TV. As another example, a student's trend feature about the cosmetic use of a muscle paralyzer, Botox, lacked the dimension a good interview can bring. When I suggested to her that she ought to talk to someone who had had a Botox injection, she said, "I don't know anyone who has." An imaginative reporter plunges into the research process and comes up with someone. Look for an appropriate chat room on the Internet, contact a plastic surgeon or locate a beauty spa that offers the procedure and ask for assistance in finding someone to interview.

Clearly, writers may interview more than one person before writing a feature. Had I anticipated the possibility of an interview as well as a review at the Augusta Barn Theatre, I would have considered who, besides Wopat, might be interviewed to learn more about him—for instance, his director, a co-star, maybe a stagehand. The student writing the Botox story decided to interview several women who'd had Botox and found some of them liked it and others didn't. The diverse views balanced the feature's perspective.

Interviews may be conducted in different ways, depending on the writer's preference and the interviewee's availability:

Face to Face

The writer meets the interviewee in person. Typically, the two sit down, sometimes over lunch or dinner. The writer jots down answers to prepared questions, allows for impromptu conversation and makes observations about the interviewee, the environment and the interviewee's relationship to that environment as well as to other people. This type of interview usually generates the most descriptive, sensory detail.

Telephone

Sometimes the writer must pose questions over the telephone. This type of interview limits the possibility for observation but can still produce good quotes and insights. Advance preparation of good questions is very important with this type of interview because long pauses become more awkward in conversation than in face-to-face meetings.

Electronic

I have never interviewed anyone electronically, but I have been interviewed by e-mail. The questions come from the writer, and the interviewee has as much time as desired to contemplate the answers. This method seems less spontaneous, although answers to questions may be more thoughtfully considered and better phrased. While telephone interviews provide limited opportunity for observations about the interviewee, the electronic method provides next to none. The most a writer might construe from this type of interview would be the interviewee's response time and typing skills.

Group

Suppose Brian Bennett and Michael Weisskopf had attempted to interview the brothers' butlers, maids and friends as a group for their story on Uday and Qusay Hussein published in Time magazine. The breadth of the writers' sources was discussed in the introduction to Chapter 4. These sources "insisted on anonymity," a fact that would have made group interviews impossible.[1] Had multiple sources been interviewed together, the possibility exists that additional insights could have been gained. Certainly, the presence of others affects what we say and the direction a conversation takes. For example, I once interviewed a Miss Indiana with her mother present. The mother inevitably brought the interview back to her daughter's wholesomeness. Had she not been present, what might we have discussed? Interviews with more than one person at a time will produce different information than private interviews with individuals. This is just a fact of human dynamics. The presence of others influences our behavior, including what we will and won't say.

Extended

Some opportunities to interview and observe may continue over an extended period of time. The writer receives permission to keep company with, sometimes even travel with, a person or group for days, months, even years. If the extended time amounts to days, a feature story most likely results. If contact continues for years, as was the case for Truman Capote as he interviewed two convicted murderers, a book results. Capote's "In Cold Blood" made journalistic history, carving his place among the founders of the "literary journalism" movement.

Whatever purpose the interview may have, solid preparation helps ensure success. Before any interview, the research possibilities detailed in Chapter 4 should have been exhausted. You're wasting everyone's time if you go off to an interview with a list of questions that could be answered by a press release or Web site.

Preparation for the Interview

Interviews should be set up in advance. If you are working for a newspaper, you usually set up interviews through a contact—perhaps an agent or the publicity director for the person, theater company or film studio. If you are freelancing for a magazine, you may need to contact the person directly or indirectly, through the person's press agent. If you are working as a corporation's or other organization's public relations writer, you will frequently interview others who work for the same company you do and will pursue internal contacts. If you work for a public relations or communications agency, you usually work through your client's contacts. Students write assignments for their professors and for campus publications. Interviews with campus administrators should be arranged through their secretaries.

When you contact someone directly to request an interview, respect that person's right to know who you are, your story angle, and intended publication plans. Not all celebrities are as willing to be interviewed as Wopat; some take a great deal of persuading, especially if they have been "burned" by the press—misquoted, misrepresented or otherwise disrespected. Likewise, don't assume that a noncelebrity necessarily has a great interest in becoming one. Lining up an interview with a well-liked maintenance man on campus may be just as difficult as getting an interview with Barbra Streisand or Marlon Brando (and those two are tough to get). Lesser-known people may be embarrassed by the idea of generating a lot of attention.

Consider where and when to interview the person. A backstage sofa is not the ideal place, unless you thrive on interruptions. If the person is agreeable, meet on the interviewee's own turf—someone's home, office or dressing room may provide insight into the person and descriptive detail that will put the reader into the interview. I once interviewed a young orchestra conductor in his bachelor apartment. When he served tea in a cup that stuck to the saucer, I understood that his intense focus on his work obscured the minutiae of daily life, such as washing dishes.

Take time to prepare a list of questions. What would your readers ask if they were in your place? What do they already know about the person or story topic? Do not marry them. Be flexible enough to depart from the list if the interview takes some delicious turn. Also, pay attention to the way you phrase your questions. A list of common question types follows.

Open-ended questions work well because they imply no particular answer. For example, in interviewing James Earl Jones for the story included in this chapter, I asked him whether he had found race to be an issue in his career. I had no expectation of how he might answer, but I expected that he would have something substantive to say. The response I received made excellent copy: "I never thought being black would be a help or a hindrance, and it hasn't been. There is nothing dramatic about being black, but there is something terribly dramatic about being human." Of course, I used that quote.

Yes/no questions can be the kiss of death. Suppose I had asked Jones, "Are you happy with 'Conan the Barbarian'?" His answer would have been simply, "No." Of course, an interviewer may follow up a yes/no question with a request for amplification: "Could you explain why?"

Leading questions guide or direct the interviewee toward a certain answer. They make us think of courtroom drama, and they are best left for lawyers. Trying to direct the interviewee to a particular answer may be unethical. For instance, I would not have asked Jones: "Weren't you embarrassed to be associated with 'Conan'?"

Instead of asking a yes/no or leading question, I asked Jones to categorize the films he'd made. The response was colorful; he put "Conan" in "the corn flakes category," his own words.

Like yes/no questions, *closed questions* generate limited responses. When I asked Jones what role he would most like to play, I expected (and received) just a name: Big Daddy (the white plantation owner in Tennessee Williams' "Cat on a Hot Tin Roof"). If the interviewer picks up on the answer, a closed question may prove to be a fruitful conversation starter. This closed question took us into the open-ended question about the impact of race on Jones' career.

In 1996 Broadway legend Carol Channing, at age 75, toured the country in her signature role as the title character in "Hello, Dolly!" She had some very bad experiences with reporters who kept asking her *loaded questions* about her age, such as, "Why are you still performing Dolly at your age?" Loaded questions convey a serious meaning and often carry emotional import. Like leading questions, loaded questions are unethical. Not only are they unfair; they also accomplish little more than making the interviewee defensive.

The Time Factor

Whether you interview someone face to face or by telephone, a time limit for the interview may be set by your contact or by the person being interviewed. Interviews can last any length of time. When I interviewed actor Colm Feore for the story in Chapter 3, I anticipated a one-hour interview. The publicity agent who set up the interview told me he'd have to leave at 4 p.m. to prepare for his evening performance. As the story reveals, the actor had other ideas. He followed me to the door, as I tried to respect the time restriction, talking all the way. As I walked, I kept writing notes. I made his eagerness to talk the angle of the story. While reporters should be respectful of advance arrangements, it is realistic to be flexible when opportunities occur.

That was the case when I landed a telephone interview with Carol Channing. My interview with this superstar of Broadway was set up by her agent, and I was informed that after 15 minutes the agent would come on the phone line and terminate the call. A phone interview carries a lot of pressure anyway because you can't see, observe or read body language of the person you're interviewing. Pauses seem much, much longer. I was so concerned about this strict enforcement of interview time that I made a sign for my office door: INTERVIEW IN PROGRESS. DO NOT DISTURB. Many times press agents initiate the interview calls, and that was the case with Carol Channing. I sat and waited, sweating bullets.

When the phone rang and the interview with Channing got underway, something happened that I could not have expected. I later found out that the time limit had been imposed because prior interviews during her tour had made her feel so uncomfortable. Reporters were asking questions about her age that she felt were sexist ("Would they ask a man questions like that?" she asked me when we discussed these interviews later). Fortunately, such questions never occurred to me. Instead, I discovered that her son, Channing Lowe, is an award-winning political cartoonist of whom she is very proud. In response to my interest, she said, "How nice of you to ask about my son." Her agent cut in on the line at the appointed time, but she said plaintively, "Could we have just a little more time?"

You have to be ready with questions you really didn't expect to have time to ask.

In the end, the conversation about her son told a lot more about Carol Channing than questions about her age would have done. The angle of the finished feature was that of all the stars I had interviewed, only Channing defied my evolving theory that celebrities are really just ordinary people who have done extraordinary things. I could find nothing at all "ordinary" about Channing. These paragraphs conclude the story:

Carol Channing doesn't take credit for Channing Lowe's gift of comic genius: "He's more erudite than I am. He's hilarious."

But the connection is so dramatic that it frames their family life like a big, black border. Channing knows her son's childhood was not ordinary, but says he has never complained.

"He never knew any other life," she explains.

While most interviews last for an hour, as the Channing interview did, an *extended (in-depth) interview* may continue over a period of two or three days. In this case, the feature writer lands the great opportunity to spend a few days interviewing someone and soaking up atmosphere for the story. These opportunities are limited by the writer's deadlines and personal schedule as well as restrictions necessitated by the schedules and privacy concerns of the person or persons interviewed.

In-depth interview opportunities may occur when a celebrity is on location for a film shoot, touring a concert or appearing in a local or regional stage production. Opportunities may also arise when a reporter follows a political campaign. Newspaper and magazine features, sometimes book-length manuscripts, benefit from prolonged access. Dominick Dunne has covered lengthy trials for Vanity Fair. In 2001 he published "Justice: Crimes, Trials, and Punishments," a book that includes his accounts of the trials of his own daughter's murderer, Claus von Bülow, Lyle and Erik Menendez ("Nightmare on Elm Drive"), and O. J. Simpson.

The advantages of extended interviews are obvious: The more access you have to a person or environment, the more deeply you can understand that person, resulting in a more insightful story. You also have the chance to ask questions later that weren't addressed in an initial interview or to request clarification and elaboration. Observing someone over a period of time also allows you to see how that person reacts to a variety of situations and people. You can make more competent judgments about someone's personality and character. The James Earl Jones feature reprinted in this chapter exhibits the extensive detail that can be gathered when the interview extends over a period of time. Extended interviews may also pose ethical questions, the subject of this chapter's Click Here on p. 84.

Even though you have more time to learn background from those you interview, it is still crucial to do the usual preliminary research. If the person is a politician, familiarize yourself with the issues. If he or she is an actor, familiarize yourself with the roles as well as the plays/films this person has done. Obtain biographical information: Where was this person born, raised, educated? Then, as always, prepare a list of interview questions and take good notes. Extended exposure makes the latter a little more challenging. You may see and hear things you can't write down until later. A good memory for words and descriptive detail always helps.

Conducting the Interview

During an interview, you will observe your interviewee. Remember that your interviewee will also observe you. Dress as professionally as possible within the realm of what is appropriate for the occasion. For an interview with a stunt pilot and a ride in his rustic World War II airplane, I realized that wearing a skirt would be inappropriate. Instead, I wore a shirt-waisted jumpsuit. To interview celebrities following the Academy Awards, reporters wear "black tie." If you're a staff writer, your publisher may have a dress code. Communication research shows that people respond most favorably to those who are dressed like them; however, keep in mind that you are asking someone to trust you to be a professional. You ought to look the part.

When you arrive, shake the interviewee's hand; provide a business card or give your name and the name of your publication, even if you think the person already knows it. Leave a superior or antagonistic attitude at home.

As the interview begins, go out of your way to make your interviewee comfortable. Think of the interviewee as a friend you've just met; this implies that you will treat the person with sensitivity. You won't have a very long interview if you begin with a loaded question. A little small talk at the start never hurts. This is a good time to let the interviewee know where and when the story will be published and what your angle will be.

Your demeanor, verbal language and body language should all communicate respect. It's a wrong assumption that in-your-face interviewing worked for Woodward and Bernstein as they investigated Watergate; even they had to finesse the hard-ball questions to get their sources to talk. Positive body language involves looking the person in the eye (but not staring), leaning forward to show active listening, keeping open posture (crossing arms and legs closes posture) and sitting close enough to show interest but not to intrude into personal space. A smile won't hurt your objectivity and can put everyone at ease. Avoid intimidation techniques: Give the interviewee time to answer and space to move. This is an interview, not an interrogation.

As you engage the interviewee in conversation, observe the person's surroundings and the way the person looks, speaks, eats. Jeff Giles provides an admirable model for observation skills. Giles' ". . . And Justice for Al," written from extended interviews with Al Pacino, appears in the June 3, 2002, issue of Newsweek. From his observations before and during a long interview in a Manhattan restaurant, Giles describes the fanfare among a restaurant's personnel when they discover Al Pacino is meeting Giles there. He describes Pacino as he looked when he first met him a week earlier: "rumpled, softly funny, slightly scattered: a favorite uncle with a giant rip in his coat." He describes an earlier ride with Pacino in the back of a New York City taxi cab. He reports what Pacino does—and doesn't—eat during the dinner interview: Pacino is on a diet and "subtly suggests you get what *he* wishes he could: meatloaf and mashed potatoes. Later, he encourages you to get dessert." He describes the gesture Pacino offers to illustrate happiness: "He puts his hands out, palms to the ground, and shakes them." He ends with a final quote in which Pacino articulates the story's angle: Pacino has patience. Giles concludes, "And, for Al Pacino, you've got all night."

Giles not only makes astute observations; he records excellent quotes.

Write your interview questions in a *stenographer's notebook*; record quotes and observations in the same book. The notebook's 6″ × 9″ paper size makes it easy to tuck inside a purse or suit coat, yet the pages are large enough to accommodate both questions and interview notes without having to flip them too often. Take notes even if you're taping an interview and always ask the interviewee's permission to tape. Some interviews take place when it's difficult to tape; for example, when you're in motion—walking, riding, moving from room to room. External noise in restaurants or heavy traffic may render tapes useless.

When I was a reporter for my campus newspaper, I decided to tape an interview with a faculty member. He was one cool prof, a bit left over from the 1960s with his blue jeans and long hair but full of witticisms and wisdom. I had no idea that a colleague's typing on the other side of his office wall would come through more clearly on the tape than the interviewee's low-pitched voice. Fortunately, the tape was a backup. Always take good notes, instead of or in addition to the taped record. And don't let good quotes get away. If you hear a profound statement while walking out at the end of the interview, stop and write it down verbatim.

Feature writers should be flexible with interview time and also with the direction the interview takes. Come prepared with an angle but let the interview take on its own life. If insightful information surfaces, listen and record. Ask the prepared questions, but be ready to go beyond them. If an answer generates a new question, for goodness sake, ask it.

Comparable etiquette and procedures apply to any interview, no matter what communication channel is used. You won't be able to shake hands and the interviewee won't be able to observe you, or visa versa, but your tone of voice and body language affect a telephone interview.

Clear the room for the phone interview; lock your door, if necessary. Ignore call waiting. I've even found it helpful to dress up a little for a phone interview; the way we dress determines how we will behave, and the goal is to be as professional as possible. Sit up straight. Free your hands by using a speaker phone, if possible; in any case, figure out a way to free your writing hand for notes. Listen very carefully. Observe everything you can about the interviewee's voice and listen for background sounds.

Even though you aren't face to face, a telephone interview is also a meeting with a new friend. It's okay to ask where the person is and what they were doing when you called. In a telephone interview with entertainer Maurice Hines, these initial questions enhanced the intimacy of what can be the most impersonal channel for an interview. The chiropractor who travels with Hines had just adjusted his back; I had just come from having my own back adjusted. This commonality created a connection that felt stronger by the time we hung up.

Ending the Interview

End the interview with these questions: "Is there anything you'd like to add? Did I forget something you feel is important?" Ask for a contact name and number in the

Click Here: A Question of Ethics

Extended interviews may occur as a series. I had interviewed Tom Wopat in a series of interviews that spanned ten years. He came not only to remember my name but also to be able to recognize me in a crowd. So he recognized me at once in June 1995 when I arrived for a reviewing assignment at the Barn Theatre in Augusta, Mich. He was at the Barn rehearsing its next play, "The Music Man," to open in two weeks.

The Barn provided press parking adjacent to the backstage doors, and Tom was just approaching the stage entrance as I got out of my car. I was surprised to see that he was angry. Recognizing me, I supposed, as a friend, he couldn't contain what had happened. He blurted it out. In a disturbing phone conversation with his agent, Tom had just learned he'd been dropped from the cast of the "Cybill" television series. The show had provided a sort of television comeback for him a decade after "The Dukes of Hazzard" had made him a star. For an actor, it was bad news. For a reporter, it was a scoop.

The next day, I phoned my editor and told her what I had learned and how I learned it. She wanted the story—after all, it was sure to go out on the AP wire. However, the information had come to me as a friend, not as a reporter. She suggested I call CBS for confirmation; that way, the story wouldn't come from an off-the-record conversation. CBS took its stance: "No comment." So I called Tom's publicist, who wondered why I'd bothered to call. "I didn't know journalists had ethics anymore," he said. "I've verified the information. Now you can print it."

None of this felt right. I wanted Tom's permission to print the story he'd unwittingly given me. The publicist called back. "Tom asks you to hold the story," he said. "It's not a done deal. If you print the story, negotiations with CBS might stop."

Ethics reflects a personal system of moral beliefs or values. For that reason, what is and is not ethical can be hard to define.

The Society of Professional Journalists, Sigma Delta Chi, publishes its Code of Ethics. All media writers should have a copy and refer to it when ethical issues arise. The code states that journalists should be truthful and accurate. I had three confirmations that the information about Tom's contract was both accurate and true. The code also states that a journalist must protect confidential sources of information; in the Wopat case, a condition of confidentiality might have been debatable, but in my mind, it was not. When it comes to ethics, it's what is in your mind and heart that matters, not some convenient technicality. It was the story I refused to write.

The story I did write, with Wopat's understanding that our conversation was all on the record, appears at the end of this chapter.

Extended interviews provide intimate, prolonged access that may raise ethical questions. What is and is not fair game? If you overhear conversations, are you free to print what was said? If you observe behavior that could be embarrassing to the person(s) involved, should you include it in your story? Journalists have to find an ethical balance between the public's right and need to know and the individual's right to privacy. Put yourself in the other person's place. Ask yourself the same questions, but answer them from that person's point of view. *Then* decide what is ethical.

To avoid problems in the first place, take a few steps to avoid misunderstandings. Be sure that you and your interviewee understand that everything in an interview is on the record and may appear in the story. If confidentiality is requested, then the interviewer has to decide whether to forego the information or to hear information considered off the record (in this case, you agree that what you hear or see will not be part of your story). If you put your notebook away, this may be interpreted as a sign that the interview is going off the record. If you're making mental notes for the story, be forthcoming about what you're doing.

As an epilogue, I have never regretted my decision to let a scoop go by in the Wopat case. What guided my decision wasn't really the Code of Ethics; it was the feeling I had in the pit of my stomach. If it makes you feel sick to do something, maybe you shouldn't be doing it.

event you need to verify a quote or specific details later. Give assurance that you will make every effort to be accurate (and then do it, of course).

Sometimes the interviewee will ask to see the story before it's published. Ethical issues must be resolved through communication during the interview and writing process; however, journalists rarely offer prior review of the story to those mentioned in it. In my experience in newspaper work, prior review was prohibited. This is not the case with public relations features, which the client or employee interviewed has the right to preview before the copy is disseminated. A public relations writer may also need a superior's final copy approval.

Remember that the interview is a meeting with a new friend. Express your thanks and best wishes accordingly. Shake hands again as you leave.

FEATURE STORY

"*James Earl Jones*"

When James Earl Jones visited the University of Notre Dame to present an evening of readings from William Shakespeare's "King Lear" on Nov. 29, 1983, I was included in the party that met Jones at the airport. Over the span of three days, I was able to interview him at an informal press conference, join him for dinner, accompany him as he spoke to Notre Dame classes, escort him around campus and attend his evening performance. The challenge with this extended interview was selecting details for the story from an embarrassment of riches—I had many pages of quotes and observations in addition to pages of preliminary research. The story was still too long for the space my editor appropriated for it; of course, it was cut. Material that did not appear in the published story has been restored in italicized print.

Dressed in a Greek fisherman's hat and a quilted parka, unzipped, over a yellow sportshirt, James Earl Jones arrived at the South Bend airport. Tall, gray of beard, Jones looked like Ernest Hemingway as he appeared on the dust jacket of "The Old Man and the Sea."

Jones was born Jan. 17, 1931, in Arkabutla, Miss. However, like Hemingway, Jones grew up in Michigan, where both were youthful hunters and fishermen. In Manistee County Jones developed a long stride he calls his Michigan walk and conquered stuttering. Unlike Hemingway, the lifelong hunter, Jones shot his last deer at the age of 14 when he decided he couldn't handle killing animals anymore.

Jones was in South Bend for a University of Notre Dame engagement, "An Evening with James Earl Jones—Readings from 'King Lear.' " He spent two days on campus, visiting classes, performing and shopping at the campus bookstore. After eye surgery this fall, Jones canceled most engagements, but he kept the Notre Dame date.

At a press conference in the library of the Morris Inn, Jones explained that it was the nature of the invitation, a chance to perform "King Lear" again, that appealed to him. It has been five years since Jones played the title role of the foolish, aging king for the New York Shakespeare Festival.

Jones has had starring roles in over 25 major films, has starred in over 40 Broadway and off-Broadway plays and created the voice of Darth Vader in the "Star Wars" trilogy. He was nominated for an Academy Award for his performance in "The Great White Hope." Most recently, his is the voice in a Chrysler commercial.

The complexity of King Lear "still overwhelms me," he said. In his portrayal of Lear, he "chose an innocence, got rid of the kingliness and the trappings of power and became just a person." If he played Lear again onstage, he would further explore these complexities. "It's more than madness," he said of the difficult role.

Jones first performed with the Shakespeare Festival in 1960 when he requested the small part of the executioner in "Measure for Measure." Since then he has played nearly all of Shakespeare's most famous characters, including both Macbeth and Macduff in "Macbeth," Claudius in "Hamlet," and Caliban in "The Tempest." Performing Shakespeare was clearly what Jones most wanted to talk about in his news conference, but questions about his performances in recent movies were inevitable.

"An actor," he pointed out, "cannot live on serious theater alone." He firmly placed his recent film credit in "Conan the Barbarian" in the "corn flakes" category. It was his turn to ask the questions. "Why did I do it?" he asked, then laughed and gestured toward his back wallet pocket. "Feel how thick that is," he said, his famous deep, resonant voice caressing each word.

"The original script for 'Conan' was interesting," he said. "I was engaged to give speeches which were parodies of Jim Jones and Hitler. I sounded off those speeches, but they didn't work and were cut. I had nothing left to do but cut off the head of Conan's mother and turn into a snake."

Next month he begins work on a new picture, "City Limits," which he calls "a punk biker movie similar to 'Bladerunner' and 'Escape From New York.' " It isn't " 'King Lear,' but it pays the bills. It's a way to stay in front of the camera." Jones prefers stage acting over film "because of the immediacy," but actors have to make tradeoffs to do the roles they really want. His father, also an actor, gave him "a sense of reality about acting. I didn't expect to be wealthy."

Actors must have had "a taste of life to be the best. To portray Juliet, for instance, you must have experienced not just love but that fever." Directors should be "guiding rather than dictating. There are times that all you need is a good play and a good cast, but if the play is complex, like Lear, you need a good director. A bad director won't do any good." Jones himself has been both an actor and a director. In 1972 he conceived and directed an all-black production of Anton Chekhov's "The Cherry Orchard" for the New York Shakespeare Festival.

Later, over dinner in the Inn's private dining room, Jones admitted that he "lusts to play Big Daddy" in Tennessee Williams' "Cat on a Hot Tin Roof." So far, he has not been able to persuade the right people that a black man could take the part of a plantation owner in the American South. Aside from that, he has not found race to be an issue in his career: "I never thought being black would be a help or a hindrance, and it hasn't been. There is nothing dramatic about being black, but there is something terribly dramatic about being human. We all have to cope with our differences. One should never think that negative thing."

When Jones started rehearsal for "Lear" at the Shakespeare Festival, he tried "to achieve what black American actors worry so much about—class, regency. It got so boring. One day I said, 'I'm tired of this.' I came in, stooped over, and started growling, being an animal, a track I could deal with. Lear is the least kingly of Shakespeare's kings. He's past dignity. He doesn't waste his energy on civilized animals. I was not ashamed that, in performance, there was a constant flood of snot and tears."

At one point during dinner Jones stopped talking abruptly, fork in hand. A woman at the table had left him temporarily speechless. "You look just like my wife," he told her. Jones married actress Cecilia Hart, one of the four women who played Desdemona to his Othello on Broadway in 1981. Reflecting on that experience, he said, "It was strange, to kill your wife every night."

Jones' recent marriage and the birth of his son, Flynn, now 11 months old, have changed his outlook on life. "We spent Thanksgiving at home, sitting in front of the fire," he recalled wistfully. As he enters his fifties, he is settling into a new home in New York State and becoming concerned about where his son will go to college. To assure his son a sound financial future, Jones' earnings from the voiceover for a Chrysler commercial have gone into Flynn's college fund.

Flynn was very much in Jones' thoughts the next day as he toured the campus. Even with its overcast November skies and barren trees, Notre Dame fascinated him. Bundled in his parka against the crisp fall air, he walked with his Michigan stride through fallen leaves to the bookstore. The man who howled and spat in his mad-dog portrayal of King Lear wandered the store aisles with infant clothes draped over his arm. He didn't mention Flynn's choice of a college as he purchased tiny Notre Dame T-shirts and sweatshirts for him, but he did verbalize a desire to introduce his son to the Catholic faith. Raised as a Methodist, Jones converted to Catholicism, inspired by a Jesuit priest he met while in the Army. Heading toward O'Shaughnessy Hall to meet with several Notre Dame classes, Jones fessed up to another reason for keeping the engagement—"I have always been curious about Notre Dame."

Jones had come to campus to do Shakespeare. In the classroom, students repeatedly asked him to do his Darth Vader voice. He concealed any disappointment he may have felt, complying with the request before turning the conversation toward Shakespeare. At one time in his life, the idea that others might someday be fascinated with his voice might have surprised him. As a child, he stuttered. During his high school days in Dublin, Mich., an English teacher took an interest in his poetry writing. One day she asked him to read a poem in front of the class, an experience that not only cured him of stuttering but also interested him in acting. He graduated from the University of Michigan as a pre-med student, but went on to become an actor who continues to write poetry, which he says can be healing: "Like prayer."

There was no stuttering when Jones took the podium in Washington Hall for his *evening with King Lear. Dressed now in a dark suit, white shirt and tie, he slid his glasses over his nose and peered at the overflow audience. Here, in the place said to be haunted by the legendary Gipper, who died from pneumonia after sleeping overnight on the hall's steps, Jones con-*

quered all ghosts. He said, "I'm gonna blow my nose, smoke, drink water, but mainly read words. I'm going to read words, Lear's words, Shakespeare's words. If you like 'Dallas' and 'Dynasty,' then 'Lear' is right up your alley. The play deals with treachery, deception, good people, bad people."

Jones' departure from the South Bend airport the next morning was less eventful. Still, Jones carried the burden of celebrity. A young lady ran up to him and asked, "Are you really James Earl Jones?" She wanted, she said, just to shake his hand.

In the boarding area, Jones said, "That was so nice. All she wanted was to shake my hand." Then she reappeared with a tablet of paper. She wanted not one but three autographs. Jones complied with grace. He remains grounded in Midwestern practicality and patience learned long ago when he played basketball and picked cherries in the summer, up in Michigan.

Chat Room

1. Richer, more extensive research materials actually make the story more difficult to write. Copious notes challenge a writer's organization skills. How is the Jones story organized? Could focus be improved through tighter selection of details or another organizational pattern?
2. What type of material did the editor cut? Are these the cuts you would have made?
3. News writing does not require transitions; feature writing does. Identify the transitional words, phrases and paragraphs in this story. Explain how they work.
4. Discuss the purpose and effectiveness of the final sentence of the story. Can you identify the allusion in the sentence? Hint: Think Hemingway.

Create

1. Jones has starred in 55 feature films, including "The Lion King" and "The Hunt for Red October." Check his Web site http://imdb.com/name/nm0000469/ for recent activities and write a new and timely lead for this story.
2. Watch "The Great White Hope." In this film, Jones plays a boxer. He knew something about that sport—his father, Robert Earl Jones, was a boxer as well as an actor. Write a paragraph contrasting Jones' physique in that film with the way it is described in my feature.
3. Read Hemingway's short story "Up in Michigan" or another of his early "Michigan" stories. Drawing on details in my feature as well as in Hemingway's story, write a vignette about James Earl Jones' Michigan youth.
4. Read Jeff Giles' ". . . And Justice for Al" in the June 3, 2002, issue of Newsweek. Using his model, interview someone at least twice and in different situations. Write an extended-interview feature structured like the Jones and Pacino stories.

"Working Man: Tom Wopat"

A chance encounter with Tom Wopat preceded the interview set up for the following story; details about the ensuing ethical dilemma appear in this chapter's Click Here on p. 84. During that chance encounter, Wopat unwittingly gave me a *scoop*, a breaking news story. Releasing that information at the time might have ended his contract negotiations with CBS. Tom's publicist was impressed with my decision not to write the story—he found it unexpectedly ethical. During the scheduled interview for this story, Tom took me to lunch, he said, on his publicist's orders. He talked about his CBS contract on the record this time. The story appeared in The South Bend Tribune on July 9, 1995.

Tom Wopat
Featured in "Working Man: Tom Wopat." Photo courtesy of the Barn Theatre, Augusta, Mich.

Since this story appeared, Tom received a Tony Award nomination for his portrayal of Frank Butler opposite Bernadette Peters in the 1999 Broadway revival of "Annie Get Your Gun"; came out with "In the Still of the Night," a CD of romantic standards, in 2000; and currently stars in the Tony Award–winning Broadway revival of "42nd Street." Indeed, Tom always works.

Tom Wopat resists the idea that he's a star.

That's history, he declared as he broke from rehearsal at the Barn Theatre for an interview over lunch at a restaurant overlooking the harbor of Gull Lake. In a crisp coral shirt and white shorts, he blended into the lake crowd, the boaters, the country club set, lunching on the white-boarded sundeck. Staring out over the blue water and the white sails of the marina boats, he looked reflective. No, he said, the idea of lake life, polka dot bikinis and Beach Boy music held no nostalgia for him.

When other baby boomers were at beach parties, he worked long hours on the family dairy farm in Lodi, Wis., where he was born on Sept. 9, 1951. Wopat is a man shaped early by a rural work ethic. He hasn't missed a show or a day's work in 30 years. He starred as Luke Duke in the long-running television series "The Dukes of Hazzard," and his life has been full since he shot the 144th and final episode.

"It's mere ego," he said of his work record. "I'd like to think nobody could replace me." Then comes the truth. "I don't want to let anybody down."

Wopat has returned for the fourth consecutive season to the theatre where he earned his Equity card in 1976. Over the years, he has appeared at the Barn Theatre in Augusta, Mich., under such manly monikers as Stone ("City of Angels"), Starbuck ("The Rainmaker"), Brick (Tennessee Williams' "Cat on a Hot Tin Roof"), Billy Bigelow ("Carousel") and Curly ("Oklahoma"). His current billing as Harold Hill in "The Music Man" sold out that production before its June 27 opening. He plays the title role in a second show, "Sweeney Todd, The Demon Barber of Fleet Street," which opens Tuesday.

In a profession plagued by unemployment, "Tom always works," said Sandy Brokaw, Wopat's West Coast publicist. Tom has starred on Broadway in "City of Angels" and the revival of "Guys and Dolls." He helped to launch two successful television series: "The Dukes of Hazzard" (1978–1985) and the new CBS hit "Cybill." Somewhere between touring with his country-rock group, the Full Moon Band, and his acting career, Wopat has found time to pursue the passions of a "normal" life—bowling, growing tomatoes, painting his house in Nashville, Tenn., and writing some "kinda country" songs.

After a tough week of wrangling with CBS over a contract for the "Cybill" show, he seemed glad to get away on a bright and sunny afternoon in his old, reliable Bronco. Although Wopat's agent is still involved in contract negotiations, Wopat expects to leave the show's regular cast but to return for several episodes next season.

For lunch, he chose a restaurant with outside seating on a deck. Here, in the Michigan countryside that has become a second home, he could speak candidly of his "Cybill" co-stars: The dog who plays Duke is "dumb as a box of rocks," Christine Baranski (Marianne on the show) "has an edge," and Alicia Witt (Zoe) "is pretty precocious, smart . . . kinda like Morticia." He doesn't share the negative views Bruce Willis has expressed about Cybill Shepherd, the new show's star and co-producer. "She's more comfortable with this show than she was with 'Moonlighting,'" Wopat said. "She's the boss. It's a whole different situation."

In the sitcom's first season, Wopat has played Jeff, Cybill's ex-husband who lives above her garage.

The strength of Wopat's personality permeated the first "Cybill" season. When Wopat was initially cast, Jeff was described just as a "stuntman." Wopat found Jeff "quite simple to identify with, pretty straight-ahead, a little girl-crazy. A good ol' guy." Wopat often wears his own clothes for the part. When a fling with an ex-wife brought Wopat another child, Jeff fathered a child by one of his ex-wives on the show.

Wopat has always longed to own "a dawg," but the mobility of show life has prohibited it. His longing for a "big mutt" resulted in the addition of a huge, obnoxious dog to the show's cast. Writers named the dog Duke, a double entendre on Wopat's Luke Duke name in "Hazzards" and an earlier "Cybill" mention of John Wayne.

Jeff is also an unashamed bungler. Almost in character, Wopat stunned the waitress who came to retrieve a ketchup bottle from our table. The top had come completely off, and the ketchup erupted volcanically. "I just squeezed it, and it blew up," he said matter-of-factly, much as Jeff would do.

Few actors who "make it" return so faithfully to their roots. Even at the height of his "Hazzards" popularity, Wopat loyally gave back to the theater that gave him his start. He has a litany of reasons: He likes the chance to star in and direct shows he might not otherwise get to do; "it keeps me in touch with what's going on theatrically so that when I get that Broadway audition, I've got some chops, I've been working onstage"; "there's a comfort level at the Barn that's conducive to good work. The social framework is there and it's consistent."

He also has input into show selection at a theater that has always had "slightly eccentric" season lineups. Parts like Brick in "Cat on a Hot Tin Roof" allow him to show his range. It's a "unique theatre experience, and that's what theatre's about. It engrosses you and, for a while, you leave your own life behind."

Wopat lobbied for years to get "Sweeney Todd" on the Barn stage. Stephen Sondheim's melodic powers demand the "legitimate" voice training Wopat received at the University of Wisconsin. Yet he would have been "disinclined to do this show without Barbara Marineau," his co-star as Mrs. Lovett, who bakes pies with Todd's victims. Because the roles are co-dependent, Wopat wanted someone of Marineau's "level of professionalism and achievement." Marineau, on leave from Broadway's "Beauty and the Beast," also starred with Wopat in "Pajama Game" (1977) and "The Robber Bridegroom" (1985).

Other "Barnies" who have gone on to strong acting careers include Dana Delany ("China Beach" and scores of films) and Patricia Wettig ("thirtysomething").

Wopat added familiar co-stars to the list of comforts that keep him coming back, but a reason he didn't list became apparent: He's at home with Midwestern morality. Mention of the O. J. Simpson murder trial and Kato Kaelin caused him to recoil: Kids deserve "more wholesome" role models, he said as we left the restaurant. He walked quickly as though trying to escape "the bad Karma" of a case that clearly disturbs him.

Wopat lives the transient show business life but hangs onto an anchor in a landlocked sea: The plate on his spit-shine '82 Bronco identifies his home state as Tennessee. He drove the Bronco, which has taken him from Nashville to Hollywood to Broadway and back many times, slowly across the sun-baked Michigan countryside.

Days later, the morning after opening "The Music Man," Wopat ducked into a Gull Lake café 15 minutes before the scheduled rehearsal for "Sweeney Todd." Like Jeff on "Cybill," he lives above someone's garage when he's in Augusta. He waited for a breakfast to go and talked about a story that had appeared in the July 4 issue of Star, a tabloid paper. The story quoted him as saying he had quit the "Cybill" show. He had never been interviewed by anyone from that publication. "They made it up," he said flatly, fighting a touch of laryngitis. "They can do that." Other quotes in the story were attributed to anonymous sources.

Wopat's celebrity status carries liabilities: loss of privacy, exploitation of a person viewed as a public commodity. He feels he must accept it: "This is the life I chose."

Whatever his talents, credentials or status with the packs in New York and Hollywood, he is like anyone else approaching middle age. "I've realized more dreams than any man has a right to expect," he said. "But, like anyone else, I have to take each day as it comes, the triumphs, the disappointments."

Then he drove away in his Bronco. This time it was raining.

Chat Room

1. Wopat is better served by the decision to honor confidentiality, but is the reader better served? Comment on the Star article.
2. Why does ethical practice vary from one person to another?
3. An underlying question threads through this story. What is that question? How does this question serve to unify diverse details? Is the question finally answered?
4. Feature writing differs from news writing in emphasis on emotion, human interest and style. What emotions might this story evoke in readers? What is the human interest angle? What is the writer's point of view? How does style reflect and support that point of view?

Create

1. Tom Wopat was nominated for a Tony Award in 1999 for his performance as Frank Butler in the Broadway production of "Annie Get Your Gun." Visit his Web site (http://wopat.com) and read reviews of his performance. List five ways a review differs from the interview feature.
2. Write a reflective piece on the subjectivity or objectivity of feature writing. How does a feature utilize the skills of both the news reporter and the fiction writer?
3. On the Internet, access the Society of Professional Journalists, Sigma Delta Chi, Code of Ethics. Write a position paper on journalistic ethics.
4. Write your own personal code of ethics.

FEATURE STORY

"Something Old, Something New"

Kate Dooley
Author of "*Something Old, Something New.*"

The stories about James Earl Jones and Tom Wopat incorporate interviews conducted to profile an individual. Not all interview situations will have this purpose, however. A person-on-the-street or local reaction feature requires the writer to interview multiple sources, perhaps strangers selected randomly, on a particular topic. As an intern at The South Bend Tribune, Kate Dooley wrote the following feature for the Sunday Lifestyle section. She sought both expert and local reaction about a trend. Here's Kate's reflection on writing this story:

"My editor and I noticed how popular the Italian charm bracelet had become and thought it would make a good story. We also noticed that the classic charm bracelet never really went out of style. I decided to talk a bit about the old bracelet and then introduce its new counterpart. I interviewed at several area shops that sell the new charms and bracelets and also interviewed random women, young and old, to show the jewelry's timelessness. I received an amazing number of responses to the article and had a really fun time researching it."

Charm bracelets are a timeless piece in a woman's jewelry collection. Worn by anyone from your grandmother to your goddaughter, the charm bracelet's popularity spans generations, and each bracelet is as unique as the individual who wears it.

Jill Ann Smith, an employee at the Spa On Colfax in South Bend, Ind., owns two charm bracelets. One was passed down to her by her mother, and the other is a gift from her best friend. "She gave me a bracelet, and I gave her one, and we add a charm every Christmas," Smith said.

The traditional bracelets, such as those Smith described, are an "O" link style with dangling charms. Recently, a new member of the charm bracelet family is attracting the attention of the youthful set as well as style-conscious women. Called the Italian charm bracelet, its wearers also choose a unique set of charms but in a sleeker design and more contemporary look.

"People of every age are buying them," said Anne Timlin, an employee of Gingerbread Cottage, University Park Mall, Mishawaka, Ind., one of the local stores that sells the new bracelet. "Everyone loves the Italian charm bracelets: little kids, teenagers, even 80-year-old women," Timlin said.

The bracelets that are attracting attention across generations vary greatly in cost. Starter bands range anywhere from $6 to $35, depending on whether the wearer chooses a brand name, such as Nomination or Zoppini, or a knockoff. Charms cost anywhere from $15 to $52 each. Initials and simpler designs cost less, while the higher-priced charms have semi-precious stones and more intricate designs. If you buy all of the charms at one time, the bracelets can be a costly endeavor. But many wear them unfinished and add to them either as they can afford to buy new charms or find charms that suit them. The less expensive alternatives are being sold at stores such as Claire's, Meijer and Kohl's. Some of their charms cost as little as two dollars.

No matter the brand, most charms are the same size and work interchangeably with other charms and basic bracelets.

Wearers acquire the Italian charms one by one, much as they did its predecessor. Charms are often significant to the wearer, reminding her of a trip or a special event in her life. Nancy Mitchell of Niles, Mich., owns the old-fashioned style bracelet as well as the new Italian bracelet. "I buy charms that have significance to my life. For instance, I have an Eiffel Tower charm from when I was in Paris and a golf charm, because I play golf," she said.

John Paskiet, an employee at Tinder Box in the mall, talked about the most unique charm he has seen. "A woman came in with a Twin Towers charm. We do not have that one yet, but she had bought it in California, so we will probably have it in a little while," he said.

Jeanne Skelton, owner of Fired and Inspired, Granger, Ind., said that she first saw the Italian charms in Los Angeles. She now sells thousands of charms in her two stores and notes that favorite charms include crosses, guardian angels and "female charms" such as cell phones, lips and high heels.

Creating the Italian charm bracelet begins with buying a starter band. The average band measures 17 links. Bands come in a variety of styles, from shiny stainless steel links to fancier versions with etched designs.

Decorated charms made of enamel, silver and gold can be purchased individually and attached to a starter band by removing a plain link and inserting the decorated one. The charm's designs encompass a wide variety of themes, ranging from initials, birthstones, flags and cartoon characters to a martini glass. Popular brands such as Zoppini and Nomination have as many as 400 charms to choose from.

The end result is a charm bracelet that is unique and enduring. New twists to the original Italian charm bracelet are charm watches and charm rings, so the possibilities are endless.

Lisa Franzke, a South Bend resident, remarked on the lasting quality of the charm bracelet: "It's a classic piece. It never goes out of style."

Create

1. Choose a topic of current controversy in your community and interview people on the street about it.
2. Create a group of three to five people and interview them together to see how their responses might differ from those interviewed individually in exercise 1.
3. Write a local reaction story. Do not write in first-person voice.
4. Make a list of other topics that would be appropriate for a local reaction story.

View

Interviews can be conducted face to face, by telephone, electronically, in a group setting or over an extended length of time. However, there are general guidelines to follow for a good interview. First, set up the interview through the appropriate person—a secretary, agent or whoever else handles the interviewee's affairs. Once you can check that off your list, prepare questions based on prior research. Open-ended questions yield the fullest, most spontaneous responses. As you begin the interview, look and act like a professional. Make your interviewee comfortable; be sensitive, not overpowering. Cover your tracks if you use a recording device—always take notes in order to avoid misquotes, misrepresentations or disrespectfulness, which relate to the code of ethics you must uphold. Record not only the words but also your observations, which will later help set the scene in the story you write. Finally, remind your interviewee that everything said is on the record, unless agreed to be otherwise. Once it's time to wrap things up, inquire whether the interviewee has any other information to add and express appreciation for the interviewee's time.

Notes

1. Brian Bennett and Michael Weisskopf, "The Sum of Two Evils," *Time* 2 June 2003:37.

📖 *Help*

electronic interview a kind of "conversation" in which the writer e-mails questions, giving the interviewee as much time as desired to contemplate the answers. This type of interview provides little possibility for observation except perhaps the time frame between Q and A and such elements as grammar, spelling and overall tone.

extended interview an interview in which the writer takes advantage of opportunities to interview and observe a person, place or situation over an extended period of time.

face-to-face interview an interview in which the writer meets the interviewee in person, jots down answers to prepared questions, allows time for impromptu conversation and makes observations about the interviewee, the environment and the interviewee's relationship to the environment and other people.

group interview an interview in which the writer interviews more than one person at a time, most commonly in a person-on-the-street or family situation.

telephone interview an interview in which the writer poses questions over the telephone. Though it does not allow for visual observation, a telephone interview can still produce good quotes and insights.

6

Attitude and Interpretation: Details

Give sorrow words.
    —*William Shakespeare*, Macbeth

We've all seen it: a red-faced toddler throws himself, kicking and screaming, onto the supermarket floor. Heads turn to stare at the abusive parent and her satanic child. We've all heard it: "Someone needs an attitude adjustment," clucks the toddler's mother.

A teenage girl marches in five minutes past curfew and, staring defiantly at her parents, slams their car keys down on the kitchen table. We've all heard it: "That girl has an attitude!"

Writers have attitudes, too, and they are just as difficult to hide.

What is attitude? The Latin root word, apere, means "to fit or fasten." Literally, the word *attitude* refers to position or posture. The American Heritage Dictionary offers a second definition: "a state of mind or feeling with regard to some matter; disposition; *an attitude of open hostility*." Certainly, the examples of the enraged toddler and defiant teenager illustrate our casual understanding of the word. The toddler's posture literally collapses as he expresses his state of mind or feeling. The teenager assumes a rigid posture to show her hostility. Attitude very often refers to a negative behavior.

Attitudes can also be positive, of course. We may have an enthusiastic attitude toward televised sports or a sympathetic attitude toward homeless animals that leads us to have a house full of them.

We may say someone "has attitude," which means the person has "pizzazz"—flair or flamboyance. For example, Oprah walks into her audience, high-fives a guest, and exclaims, "Girlfriend, you look so good today!" Oprah *has* attitude, and attitude's good to have.

Writers reveal their attitudes toward their subjects in the point of view a story takes and the interpretation of facts and observations. Even the organization of the material and the choice of quotations contribute to the overall tone of the story. Observation and description come together to give color to features. Holly James' memoir of Vidalia, Ga., her hometown, demonstrates the blending of attitude, observation and description for color.

What is the relationship of attitude to tone and interpretation? How much should feature writers allow their attitudes to show? How do they show? How much should writers interpret and how much should be left for their readers' interpretation? For answers to these questions and some feature stories with attitude, such as Holly's, read on.

Attitude and Tone

Some view attitude and tone as a kind of chicken-and-egg dilemma. It's not, really. The attitude comes first and creates the tone.

We develop attitudes early in life. The toddler in the opening scenario no doubt had been denied a box of sugar-laden cereal or a package of cookies. He expressed the emotion he felt (outrage) at this refusal, revealing his attitude through his boisterous tone. The American Heritage Dictionary defines *tone* as a "sound of distinct pitch, quality or duration." This definition fits the toddler's case—the loud, high-pitched screams and pummeling feet against the floor give us an idea of what "distinct pitch" and "duration" mean. When it comes to writing, American Heritage tells us that tone refers to a "manner of expression."

Tone provides the means to express an attitude. It is not the attitude itself.

Writers choose tonal words to express their attitudes. A list of words that convey tone appears in this chapter's Click Here on p. 101.

Interpretation

One of my favorite features appeared in the February 4, 2002, issue of The New Yorker. Nicholas Lehmann observed Senator John McCain "one Thursday morning in November," then described and interpreted, deftly, what he had observed. His feature describes the powerful senator with "powdered sugar all over his mouth," eating a doughnut "out of a white paper bag." Lehmann says that McCain "looks like a mischievous little boy who's hoping not to get nabbed for some bit of deviltry." Moments later, Lehmann gives insight into how this little devil transforms into a serious Washington leader: "McCain wiped off the powdered sugar with a napkin and rearranged himself into a more senatorial position."

Notice that Lehmann reports his observations in a way that leads the reader toward conclusions about the senator; Lehmann does not directly state the conclusions. He is careful about that. McCain's "aides wander in and out all day as if it [his office] were center stage in a drawing-room comedy." Notice the conditional (not absolute) words: "as if it were."

Lehmann doesn't say that McCain loves crowds; he observes that the "more crowded his office got, the happier and more voluble he seemed to become." Again, notice the conditional word *seemed*. The writer is taking care not to bully the reader.

Observe and describe but, as much as possible, let the reader infer the meaning of the reported details. Interpretation doesn't mean telling everything you know. Interpretation involves selecting the observed details that will "make it" to the story. Novice writers often make the mistake of including each and every observed detail, whether it adds anything to the reader's understanding or not. In many cases, a garbage dump of detail detracts from the treasures that might be hidden there. It's the writer's job to sort out the good stuff.

The Latin root for "interpret" is *interpres*, a negotiator. Think of yourself in the middle, between the observed details and the reader. For example, instead of writing, "I had no idea what a tremendous encounter I was about to experience," describe the encounter. If, indeed, you describe a "tremendous encounter," the reader will come to that conclusion. As newspaper editor Doug Toney has noted, young writers today "have a tendency to want to describe emotions rather than describe the situation."[1]

If you concretely describe, you can avoid the pitfalls of such abstract words as *beloved*, *amazing* and *loving*. The reader can see Lehmann's image of "a boyish mop of brown hair." The reader can't see "beloved." If you write that someone clutches another person to their heart and rocks them, you convey both "beloved" and "loving." You may even convey an "amazing" moment, depending upon who is clutching whom.

Show, don't tell. Give the reader the cues and let the reader decipher them. But be sure to give the cues.

Describe how someone sounds. During an interview, is a person whispering or yelling, speaking slowly or urgently? Read the details. Let's go back to the list of areas to observe (see Chapter 4) for examples of how to read some nonverbal cues.

Voices

The meaning of a person's words may be conveyed through the pitch, quality and volume of the voice as well as the pace of speech. Does the powerful president of a corporation have a child's high-pitched voice? Pitch may reveal nervousness or simply poor voice quality. A loud voice may signify excitement or simply be impolite and inappropriate. Characteristics of voice must be considered within the context of an entire situation. "The slow talker can project thoughtfulness and commitment, but talking too slowly will signal reluctance and indifference—even awkwardness."[2]

Body Language

Whole books have been written on the interpretation of body language. Read at least one of these books. You'll learn, for example, that arm crossing "has been analyzed as a 'classic defensive stance.' "[3] Although typically interpreted as a defensive or protective gesture, the arm cross may also represent "a comfortable position for relaxing the arms."

Describe body movement. In the opening scenario, I described a teenager as *marching*, a staccato walk that shows determination. Remember Molly Donnellon's feature, "Making Stone Soup," from Chapter 1? She concludes the story dramatically when she tells us that Paul rocks, literally and figuratively. Interestingly, rocking movements among victims of anxiety, like Paul, actually have a physiological basis. Stimulating the inner ear, in this case by rocking, diverts attention from the anxiety and apprehension.[4] Knowing this, we can understand why Paul can't stop rocking. This also explains why we love to ride roller coasters, rock babies and, as adults, enjoy being rocked.

An expert on nonverbal communication, observing O. J. Simpson at his criminal trial, noticed that Simpson visibly protested testimony he knew to be false but showed no visible protest when accused of murdering his wife. A good observer notices and interprets this behavior. Further explore body language on your own. "Reading People" by Jo-Ellan Dimitrius and Mark Mazzarella, a recent New York Times best seller, focuses on how to read others.

Proxemics

Interpreting your observations of people takes common sense. When you interpret the meaning of spatial relationships among people, consider the context. A group of people in a crowded space will be in intimate distance to each other, but this may not be comfortable for them. Are they bending away from each other or bending toward each other? Read nonverbal signs in clusters. Consider cultural backgrounds and proxemics. A study of World War II prisoners showed that Germans required the largest amount of personal space to maintain comfort and sanity, whereas those from Mediterranean cultures easily shared intimate space with complete strangers.

Environment and Personal Appearance

Study the home or office of the person you're writing about. In "Reading People," Dimitrius and Mazzarella argue that a "person's environment can reveal clues about her job, education, hobbies, religion, culture, marital and family status, politicial affiliation, friends, priorities, and wealth."[5] These observed details may be interpreted to "confirm, cast doubt upon, or deepen what you've already learned about someone from her personal appearance and body language—whether she is flamboyant or conservative, practical or extravagant, egotistical or humble," and so on.

A person hoping to impress the reporter might show up for an interview in a "power suit" (although Sharon Stone reversed any misconceptions about the propriety of the "power suit" in the movie "Basic Instinct"); someone with no need to impress may dress casually. Remember Colm Feore's T-shirt and blue jeans (see Chapter 3) and Tom Wopat's shorts and summer shirt (see Chapter 5)? What do these choices tell us about their levels of self-esteem?

According to a feature published a decade ago, when Cybill Shepherd opened her door to a reporter, she wore only a bathing suit. Now, that's *attitude*.

Organization of Material

In the summer of 1996, Broadway actress Marin Mazzie was coming to our area to appear in a production of "She Loves Me!" at the Augusta, Mich., Barn Theatre. I knew that she and a number of other celebrities had earned their equity cards at the regional theatre, but I didn't realize she had remained close to Barn alum Jonathan Larson. During our telephone interview, she discussed his recent death and the ensuing success of his first and last Broadway hit, "Rent." The sadness I observed in her voice told me that she still grieved for her friend.

As I wrote the story, I felt a need to stress the importance of Larson's success to American theater. A check of newspaper files indicated that the British takeover of Broadway had never been addressed for our audience nor had "Rent" been reviewed or previewed or, most likely, viewed by most of our readers. To protect the feature's nostalgic tone, I decided to write a *sidebar* to accompany the primary feature.

A *sidebar*, generally a compact 200 to 400 words, tends to be straightforward, taking an objective, newsy tone. Nevertheless, writers who master the sidebar, who turn something that short into a "dazzling" brief, may "join the exalted ranks of wordsmiths editors can depend on."[6] Writing short can be more difficult than writing long. Each word has to work overtime.

Instead of a sidebar, I wrote a *secondary feature*, a full-length story that shares the page with the *primary feature* that gets the prominent placement or headline, the top billing.

I then faced the problem of how to organize the primary feature. Should I begin with details about Mazzie's life and works? Or should I get right into the bittersweet heart—her relationship with Larson? I chose the latter. Right or wrong, for better or worse, these are the decisions writers have to make. In this case, framing the profile with Larson's life and death seemed to best convey the tone of the entire interview, to put the reader into the scene.

In retrospect, I realize that my attitude shows in the Mazzie feature; that's the reason I included it in this chapter. I felt sympathy for Mazzie. I felt sympathy for Larson, whose short, tragic life ended on the eve of the opening of his life's work, and awe for what he had done. My hope was and is that the reader will absorb the details and feel the same way I do. This strategy should evoke more powerful emotions than a forced interpretation would, for instance, had I used the words "awesome" or "amazing."

Quotations

Reporters seldom report every quote written in their notebooks. Selection of quotes may be influenced by their attitude toward the topic or the tone the piece is taking.

🖱 *Click Here: About Attitude and Tone*

What's the difference between attitude and tone?

Take a look at Holly M. James' humorous attitude toward the quirks of Vidalia, Ga., her hometown. Description creates irony (the opposite of what we expect or might feel to be appropriate). She describes one of the town's restaurants as "crawling with big belt buckles, cowboy boots, cowboy hats, and good, hard-workin', Chevy truck drivin', God-fearin' country folk." The word *crawling* conveys a sarcastic tone; "good" people might "crowd" into a restaurant, but we don't usually think of them as "crawling" in. The restaurant sits on the banks of the "mighty" Altamaha River; given the writer's self-deprecating attitude toward her Southern town and her ironic tone, we imagine that the river is not much of a river at all. She says we might see "an alligator swim by," making readers wonder what kind of swamp this is. Holly goes beyond ironic to the scornful (showing contempt or disdain) when she cautions readers to leave the restaurant before nine: "after all, only so many Budweisers can be served before a fight breaks loose."

Attitudes already exist. Our word choices reflect our attitudes and add tonality to our stories. Some words used to describe tone include ambivalent (uncertain, indecisive); caustic (sharp, satirical); condescending; cynical; derogatory; facetious (lightly joking); lugubrious (mournful, sorrowful); nostalgic; objective; ominous; remorseful; solemn; witty.

For example, I elected to partially quote Mazzie's description of her work with Larson when the two first arrived in New York City. The partial quote, "we sang whenever we could," becomes poignant when placed in the context of Larson's very short life. Another partial quote, in the feature's concluding paragraph (And the people she has met at the Barn "are my closest friends"), sets up the nostalgia of the last sentence (Except, of course, for the one who is missing).

Travel Features

The best travel features have attitude. Factual reporting and observed details create a sense of "being there" for the reader.

The travel feature does what features do best—it entertains and informs. With its combination of impressions and facts, today's travel feature won't compete with Fodor's Guide to . . . whatever city or country . . . but it will offer something the travel guide book seldom does. It will put the reader into the writer's travel experience. Ernest Hemingway's travel memoir, "A Moveable Feast," takes the reader through the streets and into the cafés of 1920s Paris. His memoir has what the best travel features aim for—quirks. Hemingway's Paris has charming cafés but it also has sewers, obnoxious people and bad weather. Holly M. James' feature about Vidalia, Ga., piles on quirk after quirk about her hometown—the alligators, the belt buckles, the real southern cookin'. Tara Blanchard's travel briefs give a dreamlike vision of two famous beaches.

In this chapter's model features, Holly and Emily Ford also show us that we needn't go far to find a travel feature topic. Consider festivals, museums, historic sites, parks, country inns and tours offered by wineries, factories, galleries and exhibits. Bruce Garrison categorizes travel writing into five types: *destination* (focus on on a particular city or country), *attraction* (focus on a place to visit, such as a winery), *service* (a how-to, such as how to buy airline tickets online), *personal experience* (focus on the writer's emotional response) and *roundup* ("a summary view of a subject by theme," such as The Top 10 Reasons to Spring Break in Key West).[7] Tara Blanchard's two feature briefs were part of a roundup of the world's best beaches.

Take advantage of vacations, travel abroad and weekend jaunts. Write about them. Share tips you wish you had known before you visited—local customs, unpublished admission charges (Emily Ford's Windy City story does this nicely), confusing airports, unsafe areas. Collect brochures, interview people, observe and record what you see and hear and feel and taste—find out where the locals eat and go there. Stray from the beaten path; explore, discover. Then recapture the experience for the reader.

"Marin Mazzie, Living Her Dream"

Marin Mazzie
Featured in "Marin Mazzie: Living Her Dream." Photo courtesy of the Barn Theatre, Augusta, Mich.

I interviewed Marin Mazzie in late July 1996. She was in New York City, appearing in "Passion" on Broadway. Although I had set out to profile Mazzie, a rising Broadway star, the core of our phone interview became Jonathan Larson, one of her best friends, who had died just six months earlier. The enormity of personal loss, both Mazzie's and the theater world's, seemed ironic when juxtaposed with Broadway's gain following the sweeping success of Larson's rock opera, "Rent."

The assigned story finally had to be divided into a primary and a secondary feature. Readers needed to understand what "Rent" meant to Broadway before they could understand Mazzie's response to his life, work and death. The following features shared the front page of the entertainment section of The South Bend Tribune on Aug. 11, 1996.

In the summers of 1980 and 1981, Marin Mazzie and Jonathan Larson served apprenticeships at the Barn Theatre in Augusta, Mich. Each had a dream: Mazzie's was specific—to originate a role on Broadway in a Stephen Sondheim musical; Larson's was grand—to revolutionize musical theater.

Some believe Larson has done this with his electrifying rock musical, the Tony Award–winning "Rent."

Mazzie has also realized her goal. In a late July telephone interview, she talked about originating the role of Clara, the 19th century Italian mistress in Sondheim's "Passion." That role on Broadway earned her a 1994 Tony nomination.

This fall she will move to Toronto to originate another role, that of Mother in the Broadway-bound musical version of E. L. Doctorow's novel "Ragtime."

For a month Mazzie will be back at the theater where she and Larson launched their careers. She opens Tuesday in "She Loves Me!" a musical she has "loved so long." Based on the Hungarian playwright Miklos Laszlo's "Parfumerie," the musical's more than 20 charming songs have the heartwarming, mid-European quality expected of the musical's collaborators, Jerry Brock and Sheldon Harnick, who also created "Fiddler on the Roof." Still, its intimacy and closely integrated score fared poorly against the brassier musicals that populated Broadway in the 1960s.

"In its beauty and simplicity, 'She Loves Me!' is much like 'Passion,' " Mazzie says. However, "Passion" remains her favorite: "It's such a special piece, and I was so very proud of it." A taped version of "Passion" will air Sept. 8 on PBS.

Mazzie found it easy to talk about her own dreams, but her "heart hurt" to talk about Larson's. "Rent" broke the chain of revivals, adaptations and spectacles that dominated Broadway for over a decade.

"The '80s was a very hard decade for Broadway," Mazzie explains. "Everyone has been waiting for a musical that could make the shift from the big spectacle. For instance, a small piece with a strong story like 'Passion' couldn't last because there was no spectacle. It's been the time of the British invasion, composers like Andrew Lloyd Weber, and all the revivals."

American composers like Larson found it almost impossible to get a break. Mazzie echoes others in believing that "Rent" may turn things around.

Indeed, Larson won three 1996 Tony Awards (Best Musical, Best Original Score, Best Book) and a Pulitzer Prize for "Rent." But the awards came posthumously. The theater world was shocked with Larson's unexpected death at age 35 from an aortic aneurysm on Jan. 25, three days after the final dress rehearsal. Larson had worked an exhausting four years to mount his play.

Mazzie and Larson had remained close friends after their apprenticeship at the Barn. With Scott Burkell, they ventured to New York and put together a singing group based on characters they had developed in Augusta. Throughout 1982, "we sang whenever we could" as Larson worked on an earlier musical, "Superbia." Larson's death cut short his career as a writer, but Mazzie says "his other music will be heard. There's a lot of it."

Her voice wavered as she admitted, "His passing really put a different take on things." Since his death, she has resolved the direction her career will take: "I want to do really great work and work with wonderful directors. Being happy is really important, being healthy and alive, and being thankful for it every day. People take things too much for granted. It's easy to be angry in this business."

She has also learned another lesson: "You can't do theater for the money." Earlier this summer, Mazzie took the part of Helen of Troy in Tina Landau's "The Trojan Women: A Love Story." She was "paid little more than carfare" for the opportunity to work with one of New York's finest contemporary directors in the East River Park Amphitheater, the original home of the New York Shakespeare Festival.

A member of the regular Barn company in 1982, Mazzie has returned to her Barn "family" in '85, '86, '90 and '91. For the next month, she will live with two of her former Western Michigan University professors, who "are like my second parents."

And the people she has met at the Barn "are my closest friends." Except, of course, for the one who is missing.

" 'Rent' Reclaims Broadway for U.S."

With its $6 million advance sales, million-dollar cast album and a bidding scrimmage over movie rights, Jonathan Larson's prize-winning "Rent" has taken out a new lease on Broadway.

Those long weary after more than a decade of British domination hope that "Rent" will restore property rights to American artists.

But how did we lose Broadway to the British anyway? Didn't we steal musical theater from the British in the first place?

Musical comedy was invented on the London stage in the 1890s, according to Peter Bailey, history professor at the University of Manitoba.

Writing for the Shaw Festival, Bailey called the British the "masters of the new global musical." Yet he finds this ironic. It was American musical comedy that swept both sides of the Atlantic in the 1920s as the world experienced "the wholesale Americanization of popular culture." London theater nearly collapsed with the advent of radio and Hollywood movies while the American stage "held up the best" in "a new generation of American musicals with jazz-influenced song and dance and spectacular singing."

Among the pivotal plays were the Gershwins' "Lady Be Good" (1926) and Jerome Kern's "Showboat" (1928). By 1929, British composer Vivian Ellis declared himself "sick of the subservience of everybody connected with our musical stage to the might of American superiority."

With the tables turned and the British bringing in unprecedented spectacle, Ellis' has become the American lament about Broadway. The list of British extravaganzas has been unrelenting, from the electronic wizardry of "Cats" to the deafening blades of "Starlight Express" to the landing of a helicopter onstage in "Miss Saigon."

American theater answered with dog-and-rope tricks in "The Will Rogers Follies" and a parade of glory days revivals, most recently a restaging of the first great American musical classic, "Showboat."

Original American musicals seldom make it to Broadway these days. If they do, their runs are short, as was Stephen Sondheim's critically acclaimed "Passion." Others stall out in regional "tryouts."

Most blame skyrocketing ticket prices created by American union scales (mounting a play in London is far cheaper, and London ticket prices reflect that fact) and lack of government support. The tremendous cost of mounting a Broadway musical causes producers to shy away from anything but the already tried-(in-London)-and-true.

However, no one seems to talk or write about the fact that this state of affairs didn't happen overnight. As early as the 1960s, young composers tried to rework the musical formulas that had seen too many decades already. But graceful, shy romances such as "She Loves Me!" just didn't sell. The American musical, now blaring through the horns of "Hello, Dolly!" and "Mame," had forgotten about understatement.

Of course, understatement wasn't characteristic of the '60s. The "real world" had become unromantic. The New York theater district suffered from urban decay. The 1970–71 Broadway season opened with 46 productions, the fewest in American theater history. Only two were financially successful.

Theater historian Glenn Litton calls the '70s the time of the "rock-environmental musical, the hit-record-album-dramatized musical, the Jewish musical, the Jesus musical, the new black musical, the history lesson musical, and the revival." A few nostalgia shows like "Grease" kept Broadway alive.

Regional theaters and national tours began to take up the slack of the late '70s. But against a backdrop of revivals ("The Pajama Game," "Gypsy," "Guys and Dolls," "Pal Joey," "My Fair Lady," "Fiddler on the Roof," "The King and I" and "Man of La Mancha"), the high-tech, tuneful British originals looked awfully good.

"Rent" now looks even better. Called "rousing," "thrilling" and "scathingly funny," the show features memorable songs ranging in style from rock to salsa to tango.

Gregory Beals, a friend of the show's creator, recalls that Larson "had to sell some books to buy a ticket to 'Dead Man Walking' " just days before the final dress rehearsal. Out of his starving-artist poverty and premature death may come a new lease on life for Broadway and American theater. Newsweek's Jack Kroll says, "There are deaths in 'Rent,' but Larson needed to balance that with rebirth."

It's about time.

FEATURE STORY

"Sweet Onion Charm"

In April 2002, while a student at Saint Mary's College, Holly M. James wrote this feature about her hometown. The class knew about Holly's beauty queen days in a Southern town famous for its unique variety of onion; we all agreed she ought to write about it, but Holly resisted. She couldn't understand what was so unusual about the place. Holly currently lives and works in Los Angeles, a long way from Vidalia, Ga. Maybe now she has figured it out. While some magazine and newspaper editors might wonder whether it would be libelous to print, especially in Vidalia, it's a good hybrid piece—in a way, it's a memoir, but it also has characteristics of both a color and travel feature.

The drive into town isn't too exciting. Georgia Highway 280 takes you through southern Georgia, along with lower South Carolina known locally as the "Low Country." The highway cuts east and west through town. On the eastern entrance, onion fields dot the horizon. On the western entrance, pine trees pop up. Signs at the east and west entrances declare, "Welcome to Vidalia, the Sweet Onion City."

Vidalia is not exactly a tourist hot spot. In fact, the only time Vidalians play host to a massive number of visitors occurs during the Vidalia Onion Festival and BRAG (Bike Ride Across Georgia). Other than that, travelers typically just drive through. To really get an insider's look into what small-town Southern life is like, they'd have to stay a few days.

The town itself boasts a booming population of about 12,000, and it continues to grow rather rapidly due to new business ventures coming into the area. Vidalia is located in Toombs County. The locals simply call it "bloody Toombs" because, many decades ago, it was known for its rather unruly residents and arguments were settled in a somewhat rogue fashion.

But times have changed.

The people of Vidalia, of the usual southern stock, are friendly. "Yes, ma'am'" and "No, sir" are typical expressions, even among the youth. Greetings shared between strangers are commonplace; should you see another driver in passing traffic, you're expected to raise a few fingers off the steering wheel to say hi. People become offended if you don't wave, or if a wave is not returned.

The city doesn't have booming nightlife, skyscrapers or anything else that would normally deem excitement. It is famous for having extremely sweet onions, in fact, so sweet that people eat them as if they were apples.

In 1931, farmer Mose Coleman found that the onions he had planted were sweet. Other people started to notice their extremely mild flavor; at first, no one wanted them. Then word of mouth got around, and people soon discovered the joy of cooking with sweet onions. Today, the onions are world famous. But when you come to Vidalia, don't expect to find some huge monument to Mose Coleman or a museum built in honor of the onions. The only thing to commemorate them, in fact, is a small plaque that sits at the eastern entrance of Hwy. 280.

What Vidalia lacks in real entertainment, it makes up for with quirks and oddities that can only be found in small Southern towns. First of all, let's talk about the food, because it really says worlds about the town. Plenty of restaurants serve local cuisine. It should be noted, though, that many restaurants carry two names: the name used by the locals and the name posted.

Oda's, located on East Main Street, has plenty of Southern wares. Oda, an extremely stout African-American woman, makes the best collard greens, cornbread and fried chicken east of the Mississippi. Those who have heart problems should stay as far away from this place as possible, because like most true Southern cooks, Oda flavors everything with bacon drippin's or grease. Those who hail from above the Mason-Dixon Line may be surprised to discover that cornbread is traditionally served as a large wafer. Don't be alarmed when Oda hands you bread with the consistency of a hard pancake instead of the expected soft, square piece.

Hilton's, or Sweat's BBQ, has the best Sunday, after-church buffet in all of Toombs County. Located on East Church Street, the Hilton offers fried chicken, fatback or side meat, chipped BBQ, cornbread, cracklin' bread, mashed potatoes, Low Country boil (complete with crawfish, sausage, corn on the cob, and fresh shrimp), fried catfish, butter beans, fried okra, black-eyed peas and squash casserole; and those are just a few of the dishes you can find there. If you recognize only one or two things in the aforementioned list as actually being food, go to the Hilton. In just an hour, you can introduce your digestive system to some of the Low Country's best cooking. Be forewarned that true, home-style Southern BBQ, whether it be chipped or served in the form of ribs, is typically pork.

Bill's Fine Baked Goods, or Bill's Doughnut Shop, is located down the street from Oda's. It's Bill's home-baked Southern desserts that set him apart—Red Velvet Cake, Pecan Pie, Lady Fingers, Divinity, and Tea-cakes or Tea cookies. I recommend the doughnut holes.

Benton Lee's Steakhouse is not only locally known but also, to some extent, is known throughout Georgia. The steakhouse is named after its owner. If you require an upscale atmosphere for your steak, avoid this place altogether. It's crawling with big belt buckles, cowboy boots, cowboy hats and good, hard-workin', Chevy truck drivin', God-fearin' country folk. Located about 15 feet away from the banks of the mighty Altamaha River, Benton Lee's serves steak on a tin platter, along with cole slaw and a baked potato. Be sure to drown the potato in butter. After so many years, Benton still overcooks the potatoes.

If you're lucky enough to get a patio seat, don't be surprised to see an alligator swim by or be too offended when a speedboat loudly passes by. A piece of advice: make sure you leave before nine. Things start to get a little heated around then; after all, only so many Budweisers can be served before a fight breaks loose.

Ethnic restaurants in the area include Natsu's Japanese steak house, very upscale with wonderful food. Even the Japanese chefs carry the local accent, so your hostess may tell you that the chef is "fixin' to be right out." Vallarte's Mexican Restaurant is also good. Not many places outside of Mexico serve food as authentic and tasty as Vallarte's. If you flirt with the servers, you just might get a free Margarita.

Those on the run who want a quick Southern lunch ought to go to Dot's Café, otherwise known to the locals as the Greasy Spoon. Dot makes a mean BLT with southern style bacon. Some say she resembles Marge Simpson, but that's merely local opinion.

Speaking of churches, Vidalia is regionally known for its abundance of worship centers. The town really likes Jesus and it shows. Drive down one street in Vidalia, and it can be bet that dollars to doughnuts, the street will be home to at least three worship centers, whether Protestant, Catholic or Jewish.

The nightlife in Vidalia is exciting simply because it resembles a soap opera. You always see someone at a bar they should not be at, like the time one of mayors showed up at Rumor's Lounge without his wife.

Kerrigan's is lots of fun, and the dance floor is huge. Live bands play weekends, and during the intermissions a DJ will spin. Music styles range from country bands to National Top 40 dance hits.

The Tree House, constructed on the side of a hill, has the unique appearance of a tree house. A tree actually runs through the bar. On Mondays, they have karaoke.

Frank's Country Club screams locals. The dance floor tiles take the shape of the St. Andrew's cross, otherwise known as the Rebel flag. Frank's carries a wide array of beer, that is, if you like all of your beer canned: Coors Light, Budweiser, Bud Light, and on some nights Icehouse. The music is loud, the beer cold, the crowd lively, and the pool tables and sticks horribly warped. If this sounds like your kind of place, it's a memorable evening, guaranteed.

As far as aesthetics go, the town is full of southern charm. Flowers surrounded by wrought iron fences dot the downtown corners. The live oak, the Georgia state tree, and pecan orchards grow all over town. At Eastertime, the drive down Achenbach Circle is breathtaking; the dogwood trees are absolutely beautiful. The churches are also pretty. Some are very ornate, while others have simple country charm. For an aerial view of the city, you might go to the Vidalia Airport and ask someone to give you a buzz of the town. Any member of the local flying club, the Sweet Onion Flyers, would be glad to do this, most likely for free.

The best time to visit is during the annual Vidalia Onion Festival. From April 26 to May 19 the festival offers pageants, rodeos, onion cookoffs, carnival rides and an onion-eating contest. You might even catch a glimpse of Miss Vidalia Onion (the South is known for its beauty queens).

Vidalia doesn't offer much in the way of traditional entertainment, but maybe that is its appeal. It's a true Southern town that has not been overrun with Yankees; as locals like to joke, "We're not Florida." Of course, fresh faces are always welcomed. To see the Low Country in its purest form, come to Vidalia. After all, where else can you eat a sweet onion?

Amelia Michalski wrote the following sidebar to accompany Holly M. James' story about Vidalia, Ga. She figured the story might generate reader interest in the annual Vidalia Onion Festival, a newsworthy event to which the feature itself could be pegged.

☞ *Sidebar: "Vidalia Onion Festival: April 10–14, 2003"*

Vidalia, Ga., is southeast of Atlanta and Macon. Just take U. S. Highway 280 to exit 67, and you're there. Upon arrival, cruise First Street to find a hotel, then get ready for the four-day festival.

Thursday

11 a.m.	The 2nd annual Onion Scramble, Rocky Creek Golf Course on Foxfire Dr. Cash prizes, $120 entry fee. Call Hal Chesser (912) 537-8805.
8 p.m.	Christian comedy with Dennis Swanberg, "America's Minister of Encouragement," Student Activism Center, Brewton-Parker College, tickets $10; http://bpc.edu.

Friday

All day	The SAM *Shortline*, a Heart of Georgia Railroad excursion train, takes visitors on hour-long rides; tickets available from participating merchants in downtown Vidalia.
11 a.m.	Visit historic Main Street Vidalia to see artists Pat Appling and Jack Fields and to view works from Savannah College of Art and Design. Author Danny New will sign copies of "Body Under Siege."
2 p.m.	Learn the history of the Vidalia onion at The Blue Marquee in Lyons, Ga. Call Tina Wheeler (912) 537-1918.
7 p.m.	Dance in the streets of downtown Vidalia to the music of The Classic Rock Allstars and the country band Out of the Blue.

Saturday

Noon	Blue Angel Air Show; advance tickets $8 at http://ticketweb.com; $10 at the gate.
3 p.m.	World Famous Onion-Eating Contest at the Arts and Crafts Festival.
7 p.m.	Chef Cynthia Creighton-Jones of Savannah's DeSoto Hilton shares her favorite Vidalia onion recipes, Southeastern Technical College; free.
8 p.m.	The World Wide Rodeo at the Rocky Creek Saddle Club Arena, Lyons, Ga.; admission $5 for ages 6–12, $10 for 13 and up.

Sunday

Wind down the weekend with the Arts and Crafts show or enjoy the carnival rides before roping up the kids and heading home.

Chat Room

1. Why do you think Holly James initially found it hard to understand that a feature about her hometown would be of interest to anyone?
2. How does Holly bring her attitude toward Vidalia into the feature? What is her attitude? Give some examples.
3. What could make a feature like Holly's more objective? What in it might concern an editor?
4. The sidebar pegs Holly's feature to the annual Vidalia Onion Festival. This makes the story timely, an aspect important for newspaper publication. Where could Holly's feature be published with the sidebar? Without it?

Create

1. Visit an art museum with your class. Divide into groups of four or five, each assigned to one painting. Observe details, discuss them, and then present your painting to the other groups.
2. Do exercise 1 but take it a step further. Interpret the details you have observed.
3. Write a feature about your hometown. Write a sidebar for your hometown feature.
4. Research "Rent." Jonathan Larson based his musical on a famous opera. Write a sidebar on this topic.

FEATURE STORY

"Chicago: City of Cheap Thrills"

"Growing up in the Chicagoland area, I have always loved the Windy City and its culture and history," says Emily Ford. "When my class was assigned a travel feature, I used it as an opportunity to learn more about it. The most difficult task was determining the angle for such a broad topic (I decided to show readers some relatively inexpensive ways to enjoy Chicago) and then organizing all the details I accumulated." The biggest challenge for Emily, as for most student travel writers, was how to resist sounding like an advertisement or travel brochure. Emily's love for her city and undergraduate writing experience paid off in an internship with Today's Chicago Woman Magazine. She writes features, oversees sections, copyedits and fact-checks—all in Windy City.

Everything seems bigger in Chicago. Skyscrapers reach endlessly into the clouds in pillars of steel and glass. The city's blocks are filled from corner to corner with impressive buildings. The sidewalks are wide, accommodating rows of professionals and their briefcases, shoppers with armfuls of bags and tourists aiming cameras. A loud hum comes from the roaring of the el, the persistent honks of taxis and the chatter of cell phone conversations. Chicago exudes the feeling of a big city.

And big city means big prices, right?

Though Chicago is a city of glitz and glamour, not all that glitters costs a bar of gold. Chicago culture, cuisine and couture can be obtained for modest prices and, sometimes, for free.

Chicago has 3,780 miles of streets, so transportation is usually needed to navigate it. However, the city offers ample public transportation, so cars and even taxis are usually not necessary. Metra provides trains from most surrounding suburbs with a round trip for approximately $7.20, which is less expensive than a movie. The CTA, Chicago Transit Authority, runs buses on a constant basis, with conveniently located stops every few blocks. For $1.50, the CTA can take you almost anywhere in the city from Lake Shore Drive to Hyde Park. Chicago also has a free weekend trolley system that runs all year long to museums, shopping and visitor attractions.

Once a cost-effective mode of transportation has been chosen, the next task is choosing a destination. Some of the most popular tourist attractions do charge admission fees, but they give some bang for the buck. Both the Sears Tower and John Hancock Center charge to enter their viewing decks but give a breathtaking view for under $10.

Navy Pier, another famed Chicago "must see," provides a special perspective on Chicago, albeit a moving one. Navy Pier's ferris wheel provides a city view at 150 feet for just $4. For those who want to stay grounded, free shows throughout the year include performances by A Piering Daily, a comedy improv group, Dock Street Stompers brass band and summer fireworks.

For those who not only want to see but also want to learn, Chicago boasts a myriad of museums. The Art Institute of Chicago is a work of art itself with its classic columns, pediments and sculpted green lions perched alongside the massive stair entrance. Located at 111 S. Michigan Ave., the Art Institute has free admission every Tuesday. Its diverse collections include works of French Impressionists, photography and European decorative arts all in the peaceful hush that pervades the marble hallways.

The Terra Museum of Art gives the impression of being a subtle storefront at 644 N. Michigan Ave. For a mere $5 suggested donation, visitors may peruse its collection of 19th and early 20th century American art, including paintings, prints and photographs.

During slow visiting periods, which usually occur from late September until January, the Field Museum, Museum of Science and Industry and the Shedd Aquarium do not charge admission on Mondays and/or Tuesdays. Guests make a day of examining dinosaur exhibits, riding in a pitch black elevator to a coal mine or observing the Seahorse Symphony.

The city's rich culture is reflected in its eclectic cuisine. To eat well and inexpensively, go for breakfast or lunch. The portions do not vary dramatically from dinner, but the prices do. Many of Chicago's restaurants offer their signature dishes for these earlier meals for under $15.

Start the day off right with a breakfast at Lou Mitchell's, 565 W. Jackson Blvd. There are no individual parties at Lou Mitchell's—a series of long tables is shared by all the customers, so corporate bankers and construction workers sit side by side. The most popular morning meal is the Spinach Special, a spinach, onion and feta cheese omelet, which comes with hash browns and homemade bread for $8.45. The 1-1/2 inch thick French toast is another tasty option for $5.75.

Heaven on Seven, with two locations at 3478 N. Clark Street and 600 N. Michigan Avenue, offers exotic Banana Foster pancakes. These flapjacks will fill you up for only $7.95.

At 17 W. Adams St., the Berghoff offers German comfort food at its best. The restaurant, a city staple, has been family owned and operated for over 100 years. It began as a café that sold Berghoff beer and became a full-scale restaurant during Prohibition. The historic building's window displays show original menus, old newspaper reviews and black-and-white photographs. Inside, large, cozy dining rooms with brass chandeliers and extensive woodwork fill four floors. Wait staff cater to professionals, families and tourists alike.

The Berghoff offers hearty German dishes such as rahm schnitzel—breaded pork cutlets covered by delectable gravy for $9.95. The meal includes side dishes of the Berghoff's famous creamed spinach, potato spatzels and all the rye bread you can eat.

The Italian Village, 71 W. Monroe St., offers more European food and ambiance. The top level offers the most traditional Italian dishes of the three floors, at the best prices. Patrons can eat in the bar or the intimate dining room with romantic dim lighting and a wall mural featuring Italian countryside. Those who would like to dine in more privacy choose booths secluded by specially sculpted cream-colored walls. For lunch, the multilayered lasagna includes salad and bread—a complete meal for just $8.95.

Spice things up a bit with Cajun cuisine, Chicago style. Heaven on Seven makes it Mardi Gras year round. Multicolored beads adorn hanging light fixtures, and the background music is lively and upbeat. A seemingly infinite number of hot sauce bottles with colorful names ("Smack My Ass and Call Me Sally Habanero") lines the walls.

Try a lunchtime cup of the renowned gumbo with spicy andouille sausage for $3.95. Order an appetizer of scrumptious Louisiana crabcakes for $6.95. Get the best classic Cajun food this side of the Bayou for a little over $10.

Dao's Thai Restaurant at 230 E. Ohio St. is the perfect dining getaway. The front door of the red building is tucked up wooden stairs so subtly that it could be missed upon first glance. The dining room is small but inviting with its red and gold décor and Asian artwork displayed on every wall. Tables align both sides of the restaurant with an eating area in the center where parties can sit on a carpeted ledge with lower tables. The atmosphere is quiet and unrushed. The polite wait staff does not hurry patrons. A heaping plate of Pad Thai noodles with chicken costs only $6.80.

If a room with a view is important, try the Signature Room on the 95th floor of the John Hancock Center. Patrons can see four states or, at the very least, the entire Loop and its multiple rooftop swimming pools. The restaurant offers a lunch buffet that changes daily but includes gourmet vegetable, meat and fish dishes for $13.95.

To truly sample classic Chicago cuisine, stop at Demon Dogs, located at 944 W. Fullerton, near DePaul University's campus. Demon Dogs is clearly a no-frills kind of place with a counter where customers stand and eat their orders. The delicious delicacies are priced well under their worth: $1.70 including french fries. Of course, to enjoy the hot dog like a true Chicagoan, leave the ketchup in the packet.

Those with a sweet tooth might stop at one of many Garrett's Popcorn shops for world-famous caramel crisp popcorn. To immediately satisfy cravings, the small and always busy shops are located all over the city. A 5 oz. bag is only $2.10. Those who prefer something to savor may sample classic Marshall Field's Frango Mints. A 1/3 lb. box of the chocolate nuggets costs $6.50.

Of course, no trip would be complete without a souvenir.

Chicago is home to such prestigious retailers as Marshall Field's, Nordstrom and Saks Fifth Avenue. The Magnificent Mile houses exclusive shops, including Coach, Kenneth Cole and Escada. Though these stores offer the best in fashion, they rarely offer the best prices.

However, Filene's Basement's two locations at 1 State St. and 830 N. Michigan Ave. sell designer items at a fraction of their regular prices. Filene's floors are packed with apparel, accessories and shoes. Racks display clothing from such famous designers as Ralph Lauren, DKNY, BCBG and Maxstudio.com.

Filene's has quite a selection of shoes, but look closely. For instance, a pair of couture-quality shoes by Givenchy may sell for about 50 percent less than usual cost. Designer handbags can also be purchased at Filene's. A leather Coach bag that retails around $250 can cost $150. TJ Maxx, 11 N. State St., offers products found in upper- and middle-scale department stores at discounted prices. A pair of camel leather shoes by BCBG Max Azria that would normally retail from $110 to $300 could be a steal at $39.99.

Marshall's, 3131 N. Clark St., is a great place to find great items at great prices. The escalator leads down to a spacious selling floor with name-brand clothes and shoes. Marshall's also provides a wide selection of bedding, framed art and other home decorations.

Chicago offers something for those on a budget and those with bulging billfolds. However, the most enjoyable aspect may be the sights and sounds of a city alive with energy, culture and history.

Walk along the Lakefront and admire the nighttime skyline with the soft glow that comes from thousands of illuminated windows. Take a stroll through the Loop, stare at timeless buildings like the Rookery or the "Chicago windows" of Carson Pirie Scott. Pause on the sidewalk and watch street performers, some in full silver garb, others swaying with saxophones.

These simple pleasures truly make Chicago a city of cheap thrills.

FEATURE STORIES

"Two Beaches"

Tara Blanchard grouped with fellow students to write a roundup of the world's best beaches. Her team focused on beaches they had already visited; locales ranged from Europe's Black Sea to Hawaii's Waikiki.

Waikiki Beach, Hawaii

Tucked away in the blue waters of the Pacific Ocean lies a place where, perhaps for a moment, reality can cease to exist.

On the southern shore of Oahu, one of the six main Hawaiian Islands, is Waikiki Beach, a two-mile strip of sand and surf. Perhaps the birthplace of Hawaiian tourism, Waikiki has annually given over four million tourists their dream vacations.

Situated next to the iconic Diamond Head Crater, Waikiki offers tourists the chance to swim, canoe, surf and snorkel in the warm waters of paradise. All along the shore, beautiful bodies bathe in the sun's rays as the scent of coconut oil fills the air. The one souvenir that no one has to pay for is a true island glow.

Kalakaua Avenue runs almost parallel to the beach, creating a strip of shops, places to eat and some of the most amazing places to stay.

The Royal Hawaiian Hotel is one of these places.

Otherwise known as "The Pink Palace of the Pacific," this hotel could not possibly be mistaken for any other. To situate yourself in the sand behind this magnificent pink structure is to place yourself in a dream come true. The beauty of the pink hotel, the beach and all who walk in its sand create a place no one would want to leave.

When the sun sets over Waikiki, stars shine from a clear sky as happy-go-lucky people trade swimsuits for club gear, walking the beach and feeling every last sip of the sweet tropical concoctions. Nearby, lovers search for a lifeguard stand or a sweeping palm tree branch under which to escape the moonlight and sneak romantic kisses.

As the sun rises again, the tide comes in along the shores of Waikiki Beach, an opportune time to claim that perfect sunbathing spot and gather a shell or two from the sand. Munch on a crisp apple from the corner ABC Store or indulge in a cone of fresh pineapple ice cream and watch as eager tourists take the day's first steps into the Pacific Ocean.

Waikiki Beach is famous for a reason.

The spirit of Aloha, the joyful sharing of life, infects anyone who swims along its shores. Whether a baby in diapers, an attractive man in a military uniform or an island native in her hula skirt, everyone falls in love with Waikiki Beach and Hawaii.

As speakers of the native Hawaiian language would say, "Pi'i ka nalu, hele ana au i kahakai!"

"Surf's up, I'm going to the beach!"

The Beaches of Cancún, Mexico

The northeastern tip of the Yucatan Peninsula was once home to the great Mayan civilization. Although little remains of the Mayan people, the land they called home can still be found by the crystal-clear waters of the Caribbean Sea, the 16-mile strip of land now known as Cancún, Mexico.

Lined with 4- and 5-star hotels, the powder white coral sands of the beach meet the Caribbean Sea as turquoise waves rush the shoreline. Such waters house the world's second largest reef barrier, making Cancún ideal for snorkeling and scuba diving.

As temperatures reach unusual heights, beachside bars such as Fat Tuesday give vacationers the opportunity to start the party early while they work on their golden tans. Wait staff must make their way through mazes of beach towels, volleyball games and beautiful bodies to offer a drink. Thirst-quenching Coronas with fresh limes or "yards" of margaritas rich with tequila are often the drinks of choice.

After all, Jose Cuervo tequila reigns as the true spirit of Cancún.

Although Cancún can please almost anyone, thousands of college co-eds flock to its beaches in search of the ultimate week of partying during their spring break vacations. As a result, many a scandalous story has been created in Cancún.

Sorry, folks, but anyone who goes to Cancún knows that what happens in Cancún stays in Cancún.

Here's a hint: hot bodies, plenty of booze, wet T-shirt contests, celebrities and MTV.

Although a wild and crazy reputation follows Cancún around the world, the beauty of its beaches, the crystal-clear waters of the Caribbean Sea and the endless glow of the sun make this part of Mexico a place where anyone can find his or her niche in the sand.

Chat Room

1. Both Emily Ford and Tara Blanchard show positive attitudes about the places they profile. However, each gives her place a different color. If you were translating their details and word choices into colors, what color would you assign each feature?
2. Word choices reflect attitudes and add tonality to stories. Refer to the list in this chapter's Click Here, then identify words and their tones in the Chicago and Beaches features.
3. Discuss places you have visited and had a terrible time (that mosquito-infested camping trip, for example). Talk about the merits of writing a travel feature about these trips.
4. If you can't visit Chicago, visit the city's Web site http://egov.cityofchicago.org. Gather and share information with your group about the city's most expensive attractions.

Create

1. Write a travel feature about a unique event or tourist site associated with your hometown. Write a sidebar to give readers additional details about places mentioned in your story.
2. If you travel during a break, note what people say and do, what they wear, where they go. Write a feature in which you capture quirks of the place. Include addresses, prices, etc.
3. Research a historical place such as 16th century London. Write a travel feature, assuming you're writing for a 16th century audience or for visitors in a time machine. Emphasize the quirks.

View

We can all see and hear attitude, but writing with attitude is a challenge. If the writer uses the wrong tone, a reader may misinterpret what the writer is trying to say. A writer creates the right tone through language. Travel features depend on the writer's observations and attitude to be engaging. For example, Holly James' southern expressions create a light, humorous tone (and give the story her personal style). Through observation and description, a writer will be able to put the reader in the story. Great writers intertwine attitude, tone, observation, interpretation and description to pull in and keep the interest of the reader. These elements make for memorable storytelling. Travel features reflect the writer's attitude toward a place and convey what makes the place unique—its quirks.

📖 *Help*

attitude a state of mind, feeling about or position on a person, place or issue.

interpretation selection of observed details for a story that leads a reader indirectly toward a certain conclusion. Interpretation does *not* mean telling everything you know about your topic.

primary feature the story that gets the most prominent placement or headline, i.e., the top billing on a newspaper page.

secondary feature a full-length story that shares the page with the primary feature but is not as prominently displayed.

sidebar a straightforward, objective piece, usually a compact 200–400 words, that accompanies a news or feature story to add additional information, background or color.

tone manner of expression. Tone provides a means to express attitude.

travel feature a story focused on the quirks of a particular place, putting the reader into the writer's experience.

Notes

1. Beverley J. Pitts et al., *The Process of Media Writing* (Boston: Allyn & Bacon, 1997) 310.
2. Julius Fast, *Body Politics* (New York: Tower Books, 1980) 148.
3. David B. Givens, "Arm-cross," *The Nonverbal Dictionary of Gestures, Signs and Body Language Cues* (Spokane: Center for Nonverbal Studies Press, 2002) 8 Sept. 2002 <http://members.aol.com/nonverbal2/diction1.htm>
4. "Balance cue," 8 Sept. 2002 <http://members.aol.com/doder1/balance1.htm>The "Balance cue" entry further explains that, after "rocking for 70 minutes . . . nursing home patients diagnosed with dementia showed up to a one-third reduction in signs of anxiety and depression."
5. Jo-Ellan Dimitrius and Mark Mazzarella, *Reading People* (New York: Ballantine Books, 1999) 77.
6. Kathryn Struckel Brogan, ed., *Writer's Market* (Cincinnati: Writer's Digest, 2003) 69.
7. Bruce Garrison, *Professional Feature Writing*, 3rd ed. (Mahwah: Lawrence Erlbaum, 1999) 300.

Part II

Feature Formats

7

The Magazine Industry

But easy writing's curst hard reading.
  —Richard Brinsley Sheridan

On days when I run errands I find myself, like most people, waiting for long periods of time. I wait at the doctor's, the dentist's and optometrist's offices, I wait at the auto shop for my car when it needs an oil change, I wait in plush chairs for my coffee at the local café, I stand in line at the grocery store, and when I get my hair done, I find myself waiting for the stylist, for the color to set and for my hair to dry.

Although these diverse places are frequented by people of all ages and backgrounds, they have one thing in common: They all provide magazines to the public to ease the boredom and monotony of waiting. Displayed on racks or tossed on tables, the magazines—full of features about the latest celebrity scandals, foolproof ways to lose weight and recipes to impress your friends—attract our attention.

As I flip through a magazine while waiting to see the doctor, I realize that even children are attracted to magazines. Many are drawn to the glossy photographs of the African jungle or European landmarks, while others are attracted to sports and fashions. As we progress through adolescence and become adults, the desire to connect to the world through periodicals grows. The direct, witty writing style can be more appealing than flowery prose. Magazine stories kindle the desire to learn more about unfamiliar topics and places.

Without the magazines that clutter our bathrooms, coffeetables, mailboxes and waiting rooms, the tabloids that pique our interest in checkout lanes and the in-flight magazines that ease the monotony of flying, we would have nothing to fill the time while waiting.[1]

Magazines are everywhere—newsstands, supermarket checkout lines, bathrooms and airplanes. Many of them, like the airline-published magazines, target very specific *niche* audiences. Niche publications have become best sellers: For example, the in-flight magazine category represents an impressive share of industry advertising

TABLE 7-1   *Top U.S. Magazines (by circulation)*

1. Modern Maturity	17,780,000
2. Reader's Digest	12,566,000
3. TV Guide	9,098,000
4. National Geographic	7,665,000
5. Better Homes & Gardens	7,601,000
6. Family Circle	4,713,000
7. Good Housekeeping	4,527,000
8. Woman's Day	4,258,000
9. Time	4,190,000
10. Ladies' Home Journal	4,101,000

Adapted from "Fact Pack, 2002 Edition," a supplement to *Advertising Age* 9 (Sept. 2002): 26 and from publishers' Web sites.

revenues and includes a dozen titles, with American Way, Sky Magazine and Southwest Airlines Spirit the top category revenue makers.[2]

Advertising Age reports regularly on the magazine industry. In addition to In-flight, the major magazine categories include Automotive, Boating & Yachting, Computers, General, Home, Men's, Metropolitan, Music/Entertainment, Outdoor & Sport, Parenthood, Photography, Science/Technology, Travel, Weeklies/Biweeklies, Women's/Fashion, Youth, Business/National, Business/Regional, National Sunday Magazines and Newspaper Sunday Magazines.[3]

Throughout the 1980s and 1990s, the magazine industry experienced rapid growth with significant increases in advertising revenues, the lifeblood of print media (See Table 7-1 for the top U.S. magazines.). The diversity of magazine types allows advertisers to target more specific groups of consumers than a medium like television or newspaper. In the first half of 2003, however, magazine ad spending increased only 6.7 percent, while newspaper saw an ad spending increase of 15.2 percent.[4]

More than ever, in an industry rocked by the 9-11 tragedies and a recession, it takes know-how and determination to break into the magazine industry. This chapter gives some basic information on the industry and how to break into it.

The Magazine

Even though some say the magazine industry faces an uncertain future, editor Tina Brown believes that "the need for depth, imagination, scope, for time and length," all the qualities magazines offer, "have never been more pressing."[5] Every medium has its Golden Age; Life brought the magazine to its peak when it provided the world's pre-television images in the mid-20th century. Yet some contemporary ad buyers "are just fine" with today's "constellation of niched titles rather than one or two heavy-hitting big names" like Life.[6]

The "ability to pinpoint specific audiences is the feature that most distinguishes"[7] magazines from other media—the diversity of magazine types makes it possible to target groups such as homemakers, brides, new parents, sports enthusiasts and computer geeks as well as groups geographically defined (for example, Midwest Living targets those who live in the Midwest) and groups generationally defined (Rolling Stone underwent a facelift for its 35th birthday to dispel the idea that it's targeted to an over-the-hill generation and not savvy young people). Gene Simmons Tongue, launched June 5, 2002, targets "a very narrow segment" of KISS fans.[8]

A magazine tends to be read at leisure, not very quickly on the run, like a newspaper. Because of its size and shape, the magazine may be carried around and read at almost any location. Magazines tend to be kept for long periods of time and "are noted for high reader involvement. Most readers spend a couple days with their new magazines" and may pass them along to others, further prolonging the magazine's life.[9]

The four-color printing process and glossy paper further differentiate magazines from newspapers and make advertising in them attractive to ad buyers. Issues may carry exclusive stories or special themes. Media buyers may work with a magazine's editorial department to arrange for a client's feature release to appear in proximity to a client's advertisement, an arrangement most newspapers shun as unethical.

The magazine has cultural power: It "creates an environment between two covers."[10]

Some magazines have become "cultural monuments." Michael Wolff, a writer for New York Magazine, cites the power of magazines to catalyze social movements: Out helped "move the gay world mainstream," Wired articulates the digital revolution, Hugh Hefner created "Playboy and a sexual revolution," Rolling Stone launched a cultural revolution, and New York Magazine changed perspectives "of who and what a city is for."[11] Furthermore, "readers are willing to pay for magazines." Wolff believes that the "silly years" of the Internet have taught the value of "what people are willing to pay for. . . . It's a great business. . . . It values a good idea higher than any enterprise."

The magazine "remains both an intellectual and glamour center of the media world."[12]

The Magazine Feature

Magazines may carry more credibility than any other media, "a reliable voice in a world that seems totally untrustworthy."[13] In the post–Sept. 11 world, magazine writers have responded to "a clarion call . . . the challenge to be more thoughtful," to respond to a general longing "for beauty and joy and humor and lightness of touch" as well as seriousness.

What distinguishes a magazine feature? Each publication has its own style and caters to its own readership. The feature writer must consider the specific needs of the publication and know its rules about freelance submissions. Even if the work is accepted for publication, the editor may make radical changes in the interest of improving story quality or fitting the publication's style.

Magazine stories generally have linear organization, with a clear beginning, middle and end. Sentences range from short to long, depending on the publication's style. For example, The New Yorker tends to publish features with complex, compound sentences, whereas Sports Illustrated for Kids prefers simpler sentences. Similarly, magazines require audience-based language. While Seventeen's stories might incorporate teenage lingo, The New Yorker's stories display a more sophisticated, adult vocabulary.

Magazine story length varies from two to twelve pages, generally longer than those stories written for and published in newspapers. This extra length allows writers the room to explore their personal styles.

Writers should expect their magazine stories to be enhanced with more plentiful, more colorful and larger visual aids (photographs, illustrations and graphics) than newspaper stories. Freelance magazine writers often provide photographs, especially if the story takes place in a part of the country far away from the magazine publisher's office.

When a story I wrote about the "Seinfeld" television show was accepted for magazine publication, the editor made it clear that I would need to provide a variety of 8-1/2″ × 11″ black-and-white glossy photographs to illustrate it. Fortunately, Castle Rock Entertainment came through with publicity photos that showed cast members and segments to which I referred in my story. I didn't have to buy the photos, but keep in mind that sometimes you do. The magazine also asked me to write the *cut lines*, captions that appear underneath the printed picture to identify who appears in it and to explain the action or context. In 25 years as a newspaper writer, I had never once had the opportunity to do that.

The Magazine Feature Market

The best resource for aspiring magazine writers is Writer's Market, a reference book that covers everything on the business of writing, from information about agents to U.S. book publishers, Canadian and international book publishers, small presses, consumer magazines, syndicates, even "The Effects of Anthrax on Writers' Submissions."[14] Writer's Market advocates good writing, knowledge of writing markets, professionalism and persistence, and it stresses the importance of the story idea. That idea must be communicated to a magazine editor in a persuasive query letter (for the scoop on the query letter, see this chapter's Click Here on the next page). Generally, a query letter should be brief (one page), initially "hook" the reader, then detail how the story will be developed, mention available photos or illustrations, expected length and delivery date.[15]

Freelance writer Susanne M. Alexander queried Massage magazine, pitched an idea for a story about massage therapist Dale Huston and received the assignment. Her query letter and the feature, as published, follow in this chapter. Study the way a professional writer pitches, then follows through with what she promised.

When looking for a place for your story, don't forget in-house publications, such as in-flight magazines. Professional freelance writer Kristin Mehus-Roe pitched a story about search and rescue dogs to Animal Watch, an in-house publication of the American Society for the Prevention of Cruelty to Animals (ASPCA). As the in-flight magazine targets airlines' passengers, the in-house magazine allows an organization

🖱 *Click Here: Crafting a Query Letter*

Kristina V. Jonusas

I wrote my first query letter for a journalism course at Saint Mary's College. My teacher, a writer for The South Bend Tribune, commented that the letter was "pretty good." After reviewing it almost a year later with my magazine writing professor and after consulting some of her research books, I quickly learned that teachers, writers and editors all have different ideas about query letters. My query letter might be what one person wants but not please another.

Although expectations differ, a query letter must include certain information. Above all, it's your chance to pitch your idea for an article or book. Obviously, this idea should fall within the publication's interest. Remember, you're trying to *sell* your idea to the editor—you want the editor to *buy* your story.

Once you've pitched your idea, briefly explain who you are, what you do, where you come from. This allows the editor to learn about you and your qualifications to write the article. Essentially, the editor wants to know that you are knowledgeable and credible to write about this area. If you have published before, this is the place to mention it. If not, don't bring it up. You want to spice up your credentials, not downgrade them.

After you explain your idea and your qualifications, indicate how you plan to develop your story. Let the editor know if you have any photographs or illustrations to go with your piece or have any plans to get them. Don't actually send visual aids. They can get lost in the stacks of letters editors receive, and you may never get them back. Finally, politely indicate when you expect to hear back from the editor.

Make sure that you have *not* included the following:

1. Pay rates—you're not bartering, you're selling.
2. A request for a sample copy—this would show that you haven't done your homework (and don't send an animal feature to an automobile magazine).
3. Photographs or illustrations, but do mention if you have them.

Once you've written your query letter, it's time to send it out. A query letter should be no longer than one page; therefore, you can fold it into thirds and send it in a #10 business envelope. Include a SASE (self-addressed, stamped envelope) folded into thirds for a #10 envelope or unfolded for a #9 envelope.

Finally, keep records of what you send and get back. This will help you stay organized in case you don't hear back from the editor within a reasonable time or the time frame you mentioned.

Hopefully, the editor will reply within the specific time you requested. Sometimes that won't happen. Don't take it personally. Editors are human and do get backlogged. This is where your records come into play. Simply mail a follow-up letter with a brief description of your query and/or a copy of the original query letter, the date you sent it, and another SASE. Then just sit back and wait.

It is important to address the letter to the specific and correct person—always do that, even if you have to telephone the publisher and ask for the appropriate editor's name and correct spelling of it.

☞ *Sidebar: Resources for Magazine Writers*

The Magazine Publishers of America Web site, http://magazine.org, provides career resources and information for those interested in working in the magazine industry. You can find a job fair, job or internship through the organization's postings. The American Society of Magazine Editors' Web site, http://asme.magazine.org, also provides career resources and information.

Freelance magazine writers should check out the online freelance directory provided by the Chicago Women in Publishing (CWIP) Web site, http://cwip.org/freelancedirectory.htm. Although headquartered in Chicago, the organization lists members from all over the United States. Each entry provides the freelancer's name, address, phone number, e-mail address and a description of writing skills and experience. CWIP also provides an online Jobvine, a source for full-time jobs and freelance work, at http://cwip.org/jobvine.htm.

Interested in working abroad? Women in Publishing, a similar organization with an online presence, comes from the United Kingdom. Click on http://wipub.org.uk for awards, international contacts and Web resources. "International contacts" span the globe—Europe, Africa, Asia, Australia and North America (the latter links to CWIP's Web site as well as Women in Publishing Boston, http://womeninpub-boston.8m.net). "Web resources" links to other professional organizations and societies, both in the United Kingdom and international.

like ASPCA to reach its supporters, contributors and members. Mehus-Roe's story, as published in Animal Watch magazine, also appears at the end of this chapter.

Writer's Market advises new writers to consider writing *sidebars* (see Chapter 6). Watch for major news stories that might generate possible topics; sidebars should be written on topics not already media saturated, related to such "hot" fields as technology, health, education, medicine and entertainment.[16]

New writers might want to consider submitting to new magazines, old magazines with new owners and small publications. "Some 800 to 900 new magazines are launched each year," according to Writer's Market, and most "don't have a vast network of writers yet."[17]

You might consider the kind of search I conducted when I was looking for a place to publish my "Seinfeld" story. I visited a library's periodicals room, wrote down all the titles related to television, went to the stacks and read the current issues. I found just the right place for my story, reviewed submission guidelines published with the editorial credits and sent off a query. It worked. The story was accepted and given a six-page spread.

In addition to checking library stacks, you might also check newsstands for possible markets for your stories. Don't doze when you're in the doctor's office or hopping on the subway. Look for those magazines that seem to be everywhere and read them.

Above all, learn to handle rejection. In the early years of my publishing career, I received stacks of rejection letters. My favorite came from a major women's magazine; the editor didn't want to buy my short story, but she did want to know if the unnamed singer who kissed me in the story was Bobby Vinton. Some writers take a practical approach and find a use for these letters, perhaps papering a wall with them or earning money by taking them to a recycling center. I filed mine. To help you understand how desperate it can feel handling rejection after rejection, let me tell you that the first check I received was for $25, and I thought I was rich. It probably didn't even reimburse me for the postage I had spent, but I didn't care. That first story payment allowed me to consider myself a professional writer.

Not every talented writer becomes a published one. It takes more than talent. It takes hard work and perseverance. It takes belief in yourself and the footwork to track down the right places to submit.

After you've read Susanne M. Alexander's magazine feature, take a look at Ann Basinski's personality profile of Alexander. The professional magazine writer talks about her own background and offers advice to aspiring writers that goes beyond a textbook chapter.

FEATURE STORY

"Sports Massage Is His Passion"

As her query letter (Figure 7-1) explains, Susanne M. Alexander met Dale Huston while he was working a wellness cruise. Impressed with his expertise and energy, she asked him to work with her on a story. A professional freelancer who clearly works with efficiency, Alexander assigned a photographer to shadow Huston at two sports events. From the photographs and telephone interviews with both Huston and some of the athletes, Alexander researched and wrote the following story. It appeared in Massage magazine, issue 98, July–Aug. 2002. A niche publication that describes its purpose as "exploring today's touch therapies," the magazine appears bimonthly, with 25 percent of its stories—about massage and other touch therapies—written by freelancers. The publisher advises freelancers to study a few back issues before sending a query.

Susanne M. Alexander
Author of "Sports Massage Is His Passion." Photo courtesy of BCR Studios, Euclid, Ohio.

Young people crowd around sports-massage therapist Dale Huston, listening and learning. He's talking nonstop while massaging and stretching a runner on his table. Other athletes watch and wait for their turns.

Huston's no-nonsense voice delivers short quick sentences on stretching techniques, how to prevent injury, the importance of flushing the lymphatic system with water, and what part of the body he is working on and why. He often jokes with the athletes to distract them from pain or anxiety about upcoming events. "I'm not quiet when you're on my table," says Huston. "I'm going to educate you." It's a natural part of his style.

Sports massage is not generally the world of scented lotion, candlelight and soft music, but rather quick sessions on a portable table set up near crowded athletic events. It's sweaty, noisy, chaotic and fun. "People don't come to me for relaxation," says Huston, "but when the body is flushed of its toxins, relaxation is the ultimate effect."

Huston's popular and well known among athletes, high school and college coaches, and weekend warriors in Indiana and Michigan. He started out by contacting high school, college and amateur-level coaches and volunteering at their sports events near his home in Mishawaka, Ind. He came before the events, set up his table and got the athletes ready to compete. Competitors lined up at his table for more work in between events, and again at the end.

Coaches became comfortable with him and began to trust him with the well-being of their athletes. Athletes appreciated the education he gave them about their own bodies. Word of mouth among the athletes spread his positive reputation. Huston also advertises his services in local and statewide newspapers in Indiana, and accepts speaking engagements regularly. "I'm a good talker," he says. "I share myself at every opportunity. Even telemarketers who call my house hear about massage!"

Gradually Huston's clientele has built to include individuals and teams in a range of sports including track and field, golf, handball, tennis, hockey, basketball, rowing, swimming, bowling and kayaking.

Huston's willingness to put in long hours, coupled with the effectiveness of his massages, began to be noticed. John Millar is the women's hurdles development coordinator for the United States Track & Field team in South Bend. His program has hired Huston to work on athletes as preventive maintenance to keep them loose and injury free. "Massage is important, as it allows athletes to train at higher levels. There's always a fine line between training too much or too little," says Millar. "Massage allows us to get as close to that line as possible without injury. Dale Huston has been a tremendous asset to us . . . we have found not only have our athletes been able to increase the intensity of their training and worked harder and longer, but also reduced the risk of injury."

Huston believes that sports massage therapists require a deeper level of understanding of muscle functioning than other therapists because they advise the athletes on proper care of their muscles. Athletes, in turn, also tend to be more knowledgeable about their bodies than other clients and are willing to learn new stretches and exercises to stay limber, he says.

Huston says his special touch with clients comes from his earliest training at the South Bend YMCA in 1969 with Andy Conners, a skilled therapist who was blind. Conners died 18 months later following eye surgery. The idea of specializing in sports massage was planted at the YMCA, where Huston worked days part time as a massage therapist and helped at such events as a national handball tournament with athletes from all over the world. At night, he was a steel melter with Wheelabrator-Frye Corporation.

In 1992, Huston attended the Lewis School of Massage in Hobart, Ind., where he was inspired to start doing sports massage full time. His commitment to ongoing learning has put him in a string of training sessions and seminars ever since. He is founder, past president and current community events coordinator of the Michiana Massage Therapists Association, a regional association for the border area between Michigan and Indiana. He serves as first vice president and membership chair of the Indiana Chapter of the American Massage Therapy Association. He also coordinates the chapter's massage emergency response team for the state of Indiana.

Although he's 58 and getting gray around the edges, Huston does not intend to give up his 60-hour work weeks anytime soon, or perhaps ever. "I have such a feeling for humanity," he says. "I love this and I'm not going to give it up. God's given me the health, stamina and courage to do it, and I'm not going to stop."

FIGURE 7-1 *Professional Query Letter*

SUSANNE M. ALEXANDER

P.O. Box 23085
Cleveland, OH 44123

Phone: (216) 731-7799
FAX: (216) 731-7950
E-mail: Susanne99@aol.com

February 2, 2000

Karen Menehan, Editor
Massage Magazine
1315 West Mallon Avenue
Spokane, WA 99201

Dear Ms. Menehan:

Query: Sports Massage Promotes Peak Performance

"I don't play games when it's about an athlete's health and well-being," said Dale Huston, sports injury massage specialist. "The sooner I get them rehabbed, the sooner they are on their feet again ready to compete." Direct, strong and powerfully committed to the benefits of massage, Huston doesn't pull punches. He does his best for his clients, but he equally expects them to do what he educates them to do. With Huston's skills, he is able to significantly shorten injury recovery time, so most of his clients are glad to have him on their side.

The American Massage Therapy Association (AMTA) recommends sports massage as a key contributor to high performance, as important a factor as a carefully monitored diet. It reduces the heart rate and blood pressure, increases blood circulation and lymph flow, reduces muscle tension/spasm, improves range of motion and helps relieve pain, all of which enhances full body health and readiness.

Background: Dale Huston will be doing his method of sports massage this year at the Amateur Athletic Union, Youth Indoor Nationals, track and field, in Merrillville, Indiana, and the NCAA indoor track and field event in Indianapolis. He has also done massage for gymnastics, swimming and even a hockey team (Detroit Red Wings). He has worked with four Olympic athletes and is awaiting word about going to Sydney in 2000. A unique point about Huston—he was trained by someone who was blind.

This story could include:
- Methods Mr. Huston recommends for stretching
- Immediate response techniques after injury
- Massage techniques especially designed for athletes
- How to tell when an athlete should stop competing temporarily or permanently
- Interviews with athletes that Mr. Huston has worked on and the results
- Interview with a representative of the AMTA about the value of sports massage

I had the opportunity in November to observe Mr. Huston at a wellness seminar that featured massage aboard the Carnival Paradise (Don Alsbro's Wellness Group). He has agreed to be interviewed for this story. I've written a number of health-related articles for The (Cleveland) Plain Dealer and The Shifting Times. My résumé and clips are enclosed. This is an interesting story to write, and it will contribute to greater knowledge for massage therapists. I look forward to hearing from you.

Sincerely,

Susanne M. Alexander
Freelance Journalist

Chat Room

1. Susanne Alexander makes some promises in her query letter. Does she keep them all?
2. Discuss ways Alexander's query letter resembles a feature in language and structure.
3. What is Alexander's angle in the Huston feature?
4. Who is the audience for the Huston feature? Will Alexander's angle and the material she selected for the story appeal to this audience?

Create

1. Consult Writer's Market. Find a publication you believe might be interested in one of your feature story ideas. Write a query letter and send it to the magazine.
2. In Writer's Market, identify a consumer-oriented magazine for which you would like to write. Come up with a story idea, then write a query letter proposing it.
3. Take a field trip to your campus library's periodical center. Search for magazines that might publish student-written features. Write a query letter proposing a story idea to one of them.
4. Look through magazines and find a feature written for a niche audience, perhaps a story in a bridal magazine. Rewrite several paragraphs so that the story would be appropriate for a different audience—for example, wedding planners instead of brides.

FEATURE STORY

"A Writer's Words"

Student Ann Basinski wrote the following interview feature "on assignment" just for this book. According to Ann, "Before I called Ms. Alexander, I was a little nervous. I had never interviewed anyone over the telephone before. In fact, I had never interviewed anyone I didn't know. After some light chatter with Ms. Alexander, however, I was able to ease into the questions that I had for her. I think the key to interviewing is to feel comfortable with the person. Once I'd made a personal connection, I could successfully complete the interview. The phone interview was a great experience because it helped me break out of my comfort zone, forcing me to ask personal questions of someone I did not know."

"I will never not speak up again," Susanne M. Alexander remarks as she recounts an incident from her early days as a writer.

Several years ago, before Alexander became a successful freelance journalist and president of ClariComm Group, a company that specializes in "clear communications and bright ideas" for businesses, she was asked to write a story for a local Cleveland newspaper. The editor of the paper wanted Alexander to write a 6,000-word story on baseball. This assignment would not have been difficult had it not been given during autumn, after the baseball season was over.

Eager to please her new boss, Alexander set out to complete her assignment. After a lot of hard work, and not too much luck, she was able to finish the article. But, after presenting the article to her editor, she was forced to rewrite it because, according to her editor, there was "not enough baseball" in the story.

At that point in her career, Alexander realized that editors do not have complete control over a writer. She could no longer put her editors on a pedestal. She recognized that writers and editors must have fluid and equal communication. If she had questions about an assignment, she needed to articulate them. Alexander realized that as a writer she had a voice and that communication with her editors should not be blocked.

"We're partners in the process," Alexander said enthusiastically during a recent phone interview, as she described her current relationship with her editors. With poise and confidence, Alexander said she now appreciates input from editors and has good relationships with them.

Her enthusiasm for writing stems from her freedom to express herself; to do so effectively, Alexander must not be hindered by a poor writer-editor relationship.

Alexander candidly admits that not all writer-editor relationships result in positive partnerships. She remembers an incident when an editor added quoted remarks to one of her interview pieces. Alexander laughed and added that she "doesn't write for her anymore."

Before freelancing and ClariComm, Alexander and her husband had started a family in Cleveland. She returned to school and became a communications major with an emphasis in business at Baldwin Wallace College in Berea, Ohio. At Baldwin Wallace, she took part in an assessment to prior learning course, which earned her 45 credits in one year. She took other writing courses and soon learned that she preferred nonfiction to fiction.

After graduating cum laude from Baldwin Wallace in 1996, Alexander first submitted articles to editors of local civic organizations and quickly started getting published. One article, published around Christmas and about a reunion between a mother and daughter, caught the eye of an editor for a daily Cleveland newspaper. Soon after, Alexander began writing features for the health section of that paper.

Now Alexander writes on a fairly regular basis for Newsweek Japan. She faxes articles on U.S. business to New York, where they are edited and sent to Tokyo to be translated into Japanese. Other publications that have printed Alexander's articles include Ziff Davis Smart Business, Massage magazine, Writer's Digest, Pages, The Cleveland Plain Dealer, Catalyst: For Cleveland Schools, Crain's Cleveland Business, Over the Back Fence, and The American Baha'i. In 2001, she co-authored "The Preparation for Marriage Workbook" and has written and edited several other books. Alexander was also a staff writer and associate editor for The Shifting Times, an Ohio monthly publication.

Recently, Alexander has found it difficult to find story ideas about good things happening in the world of American business. With the downfall of the economy and the corruption and scandals now surfacing in American businesses, Alexander, with a slightly downcast voice, talks about a "downward spiral of bad news."

"[When I interview,] I'm not digging up dirt on people in business situations," Alexander says. "I write about innovative people who are doing fascinating things" related to American business.

Alexander thinks that interviewing people is "great fun." She says animatedly, "It's always a bit of a game to find out how quickly I can build a relationship and connection with the person who I am interviewing." Alexander likes getting the chance to meet new people and to "find out what makes them tick."

Unfortunately, Alexander has noticed that some businesspeople are overly prepped for interviews by their public relations people. Instead of giving a true sense of who these people are, Alexander has found that many corporate leaders' interviews are stilted and artificial.

Alexander has a passion for learning about new things. Writing allows her to explore that passion, especially when she is asked to write about things to which she has had little or no prior exposure. Alexander compares writing to going to school. Beginning as a freelance writer, Alexander says with confidence, "I set out to educate myself." She attended writing conferences and workshops, picking up mentors to help her with her writing.

For aspiring writers, she advocates having a strong support system. Mentors and encouraging editors provide positive support in a career where even the best writers receive rejection letters and negative criticism. "You have to be willing to make mistakes and look at each step along the way as a learning experience," Alexander says enthusiastically.

FIGURE 7-2 *Another Query Model*

SUSANNE M. ALEXANDER

P.O. Box 23085
Euclid, OH 44123
www.ClariComm.com
Phone: (216) 383-9943
FAX: (216) 383-9953
E-mail: Susanne99@aol.com
Member, American Society of Journalists & Authors
Cleveland Board Member, Society of Professional Journalists

June 7, 2001

Edward S. McFadden, Senior Editor
Reader's Digest
1730 Rhode Island Ave., NW, Suite 406
Washington, DC 20036

Dear Ed:

I enjoyed hearing you speak and meeting you briefly at the ASJA conference in New York on May 19. I also appreciated your input about the conference when I called to interview you for the Writer's Digest article I'm working on. As we discussed during our conversation, here is a story proposal that I believe will work well for Reader's Digest:

Fear, excitement and shyness chase through their eyes, as fast as the clouds in windy Chicago where they've landed or a stray bullet in their native Belfast, Northern Ireland. At 11 and 12, they are wide eyed and curious, but they also arrive with the baggage of all their country's conflicts and prejudices.

"They are so excited to have this adventure, completely mesmerized. They've never been on a plane, never been out of Belfast, and they need a taste of normal childhood," says Diane O'Connor. And bringing children mentally, physically and spiritually out of the culture of Belfast is just what she and her Dublin-born husband Rob have been doing for 20 years.

"There's a cycle of hatred there that goes back more than 300 years. Our goal was to break that cycle," says Rob O'Connor. "We believe with all our hearts that this is a calling. Diane and I and the children's parents view these children as the hope for the future."

What these children know are stressed families, unemployed fathers, neighborhoods with tall walls dividing Catholics and Protestants, and the violence that has touched every family. These participants in the Northern Ireland Children's Project have the chance to expand their worlds and change their futures. For five weeks each summer, up to 200 children stay with stable families in the Chicago area. Social and sports events are set up to mix the Catholic and Protestant participants.

On the trip to Chicago, the plane is rigidly self-segregated. On the plane home, they are completely mixed. Rob and Diane say the children have discovered that they look like each other, dress like each other and play soccer like each other. They are no longer sitting on their own sides of the aisle—segregation has been transformed into friendships. "At 11 and 12, the children are old enough to travel and leave home, but their minds aren't so set," says Diane. "By 15 or 16, they are pretty molded, and their minds are set anti-Catholic or anti-Protestant."

The Doherty family in Indiana, a host family for 20 years, recently heard from a man who had stayed with them as a boy in the first year of the program. They had not heard from him since that time. He found their phone number in a drawer and called to say the experience of living with them had touched and changed him. He wanted the Dohertys to know that the visit had given him a vision of the family he wanted to have and that he was now in a stable marriage with four children of his own.

Rob and Diane O'Connor have agreed to be interviewed further and to arrange access to other people important to the story. A story about this project could include:

- The Chicago 20th anniversary reunion party July 8, 2001, in Gaelic Park
- Interviews with host families and those who participated at ages 11-12 or those who have stayed in the program in Ireland and come back at age 15
- The range of activities for the children in Chicago and Ireland
- Challenges (and joys) facing coordinators, host families and children
- How people can get involved in the United States and Ireland
- The O'Connors' spiritual motivations and commitment

In addition to being a member of ASJA, I'm also a Cleveland board member of the Society of Professional Journalists. I've had around 200 stories published in The (Cleveland) Plain Dealer, Crain's Cleveland Business, Ad Astra (National Space Society), Black Diaspora, and Over the Back Fence, among others. I'm currently working on assignments for Newsweek, Proud, Kaleidoscope, and Women As Managers. Clips are enclosed.

I look forward to hearing from and working with you.

Sincerely,

Susanne M. Alexander

"Disaster Search and Rescue Dogs"

Kristin Mehus-Roe is a freelance writer and editor based in Long Beach, Calif. This story originally appeared in the Fall 2002 issue of Animal Watch, a niche magazine published by the American Society for the Prevention of Cruelty to Animals (ASPCA) and "Dedicated to the Protection of All Animals," and is also reprinted in Mehus-Roe's book "Working Dogs" (Bowtie Press 2003). The story's subtitle invited readers to learn more about the lives of rescue dogs: "They make it look easy and they do it for fun. But behind a certified disaster dog is hundreds of miles, thousands of dollars and continuous training."

Manny, a border collie, leaps across jutting stone and twisted metal on a large rubble pile. He zeroes in on a wooden pallet wedged between chunks of cement and begins to bark. Hidden underneath is Debra Tosch, a search dog handler and the executive director of the National Disaster Search Dog Foundations (NDSDF) in Ojai, Calif. Tosch doesn't move until Manny's handler, Ron Weckbacher, reaches them. Then, Tosch extends a hand from her hiding place to offer Manny his reward—a tug toy. "Good Manny, good dog!" cheer both handlers. Manny grasps the toy and pulls, dancing and whining in excitement. To dogs like Manny, search and rescue is a game, and that game is their life.

On a beautiful, quiet day in Southern California, it's hard to believe that this game Manny plays so avidly is all in preparation for other, less calm days, in less calm places. Days like the one last September when Weckbacher and Manny, and Tosch and her black Lab, Abby, were deployed with 11 other NDSDF teams to search for victims buried beneath the remains of the World Trade Center in New York City. Like all members of NDSDF, Tosch and Weckbacher train with their dogs week in and week out so that they'll be ready for that terrifying moment when the sky falls in.

Name of the Game

The dogs used for disaster search and rescue (SAR), also called urban search and rescue, use their noses to find living victims trapped in the crannies and voids created when a building collapses due to an earthquake, hurricane or explosion. Other SAR dogs are trained in wilderness, avalanche or water searches—each type of SAR demands specific training. Disaster dogs must be able to focus on their search while navigating large piles of shifting rubble and contending with distractions that may include other search dogs and people, and the presence of cadavers.

The seed for the NDSDF was sown when retired physical education teacher and certified dog handler Wilma Melville was deployed in the aftermath of the Oklahoma City bombing. In 1995, Melville and her dog, Murphy, were among only 15 Federal Emergency Management Agency (FEMA)–certified disaster dogs in the United States. Most of the dog teams sent to Oklahoma City were not FEMA certified, meaning that they hadn't passed either the Basic or Advanced certification protocols that attest to their level of proficiency. Melville acknowledges that there's no way to say for certain that additional victims would have been found alive had there been more certified SAR dogs in Oklahoma City. However, one thing she will say: "People trapped in rubble have a narrow window of time open for their survival. The sooner they're rescued, the better their chances."

Before Oklahoma City, Melville was much like any other volunteer SAR handler in the United States. She had retired in Ojai, California, had a dog, Topah, whom she enjoyed teaching new things, and decided to pursue wilderness SAR as a hobby. She joined an SAR group, but they met sporadically and, after several years, she and her dog hadn't advanced far enough to participate in an actual wilderness search. "I went through two fairly typical years of trying to find out how to be taught properly," says Melville.

It wasn't until Melville switched to disaster SAR that she met Pluis Davern, a gun-dog trainer with experience training dogs for disaster search. "Pluis raised the bar," says Melville. "She said, 'This is what any dog can do, and this is what exceptional dogs can do.' " Although Topah was "not a bad dog," Melville started over with a black Lab, Murphy, a dog she chose specifically for her high drive.

Under Davern's tutelage, the team came along quickly. "We reached Advanced certification with virtually no problem," says Melville. "I found that there are three necessary ingredients for success: a handler who is committed and willing to learn; a dog with all of the right characteristics; and the proper person to teach them both. What a revelation!"

After Oklahoma, Melville knew that if another disaster struck, there would still not be enough dogs for the job. FEMA estimates that at least 300 Advanced-certified teams are needed to cope with a large-scale disaster. Seven years later, the organization inspired by Melville's despair over Oklahoma City has produced 26 FEMA-certified SAR teams. Of the four dozen teams at the World Trade Center, 13 were trained by NDSDF.

Ingredients for Success

Disaster dogs must have high drive and great noses, and also be well socialized, obedient and agile, so Melville looks for dogs with these attributes. She chooses golden and Labrador retrievers, border collies and mixes thereof. She opts to use shelter/rescue dogs because they often have the qualities needed for SAR work. Ironically, the energy and drive that make them difficult house pets are ideal qualities in a search dog.

Because SAR handlers must be willing to train endlessly for an event that will probably never happen in their dog's lifetime, Melville works California's Office of Emergency Systems to recruit firefighters as handlers. Firefighters are accustomed to endless training and already have flexible schedules, a connection to rescue and the emotional strength to face a disaster site.

When funding is available, Melville selects dogs from 10 to 18 months of age and places them for six months with Pluis Davern at her Gilroy, Calif., training facility. After that, the dogs are joined by their new handlers, who have received a week's training prior to meeting them. The handlers stay in Gilroy for a week to learn how to work with their dogs. During the next three months, the handlers return to Gilroy frequently to address training issues that arise.

Throughout their careers, NDSDF teams train daily on agility and obedience and twice weekly on rubble piles. During the first year, the handlers log many hours of driving time, exposing their dogs to diverse search areas from San Diego to Sacramento. Most handlers estimate that they spend $5,000 per year on travel, equipment and dog care. Although NDSDF covers the initial six-month training for firefight teams—civilians pay for the training themselves—it does not cover these incidentals.

After developing the program for almost seven years, Melville and the foundation's handlers and supporters felt they were on the right track, but it took the attack on the World Trade Center for them to realize just how far they had come.

The Day the Sky Fell

Seth Peacock recalls his arrival at Ground Zero. "I was the last one out of the can. As soon as I go out, some guy spotted me and said, 'Hey! We need dogs down here!' " Although stunned by the size of the rubble field, Peacock knew what his first priority must be. "You're walking into this rubble and you don't know what's underneath and there are hazards overhead," he recalls. "I was in complete awe, but I also had to think, now what's my next move, what is my dog doing, and how do I stay alive?"

Advanced-certified Pup Dog, meanwhile, was straining at the leash.

Tosch and Weckbacher and their dogs arrived days later but had similar first impressions. Says Weckbacher, "When you get there and the towers are no longer part of the skyline and all you see is smoke and lights coming from where the Trade Center once was . . . you feel like you're in a dream." Manny and Abby, like Pup Dog, were raring to go. Endless repetition of hundreds of sites had conditioned the NDSDF dogs to any situation. "When I released Abby for the search," says Tosch, "her attitude was 'Yes! A new playground!' "

Abby's agility on the rubble pile impressed even her handler. "We learned very quickly that we're absolutely doing something right [at NDSDF]," says Tosch. "Our dogs just went out there and had no problem negotiating the rubble, crossing six-inch I-beams over a 40-foot drop, with the beams warped and metal moving."

"Not all the dogs in New York were anywhere near the level of the foundation dogs," says Peacock. Inexperienced dogs are at risk of injury, he says, because they tense up, lose their footing and fall. Also, constant training on rubble piles toughened the NDSDF dogs' pads, so they did not need to wear protective booties that can make a dog less sure footed.

"It was a judgment call every time you went on a search," says Tosch. "Do I put booties on her, or don't I? I always decided she was better off without them so she could have that footing. I was more worried about her falling 40 feet than getting a cut.

"We were very careful," Tosch continues. "We didn't send them into hot areas, we didn't send them into voids that hadn't been checked by structural engineers and hazmat [hazardous materials] specialists. I was constantly watching and worrying—it's our job to watch them very closely." Among the 13 NDSDF teams at Ground Zero, there were no injuries, not even a cut paw.

"I don't want to undermine anybody's commitment or training," says Weckbacher, "but there were dogs [in New York] who really should not have been there. To have a dog who is untrained working in a disaster is unfair to the dog. Not only do we train these dogs, but we also need to be advocates for them. They have limitations, and it's important to understand what they are and not ask them to do things they're not capable of." SAR dogs trained for wilderness or tracking simply haven't been trained to deal with the complexities of a disaster site.

Peacock adds that inexperienced dogs also adversely affect the search process. "Even if the dog goes over that area, we can't say that it's been cleared." What a dog doesn't find is often as important as what it does. Once a FEMA-certified dog clears an area, the searchers can move on.

Although Peacock was on the pile when two of the last living victims, firefighters, were pulled from the rubble, no living victims were found after the night of Sept. 13. As a result, the search became frustrating for both handlers and dogs. To keep the dogs motivated, the NDSDF handlers made a point of doing "runaways" each night, hiding a handler with a tug toy and sending the dog to search. "They don't realize this is a job," explain Weckbacher, "they think it's fun. Not being rewarded, not getting their paycheck, they start to get a little bit down."

Playing for Keeps

NDSDF dogs are trained to find live victims, not cadavers, but most search dogs have passive, untrained responses to cadaver smell. Some dogs will slow and wag their tails slowly, or move their head sideways, as if trying to figure out the scent. Other dogs will urinate on or try to roll in an area where there is cadaver smell. Highly experienced handlers know their dogs well enough to reliably interpret this behavior. "Our job is to understand our dogs and what they're trying to tell us," say Weckbacher. At Ground Zero, NDSDF handlers let workers know when their dogs indicated trapped bodies but then moved on in their search for living victims.

The NDSDF handlers were awed by the ability their dogs showed, knowing that if there had been living victims, they would have found them. "I couldn't have asked anything more of her," says Tosch of Abby. "Everything I did ask her to do, she gave so willingly." Weckbacher adds, "I've always approached this with the idea that if it were me or my wife, or if it were my son, I'd want the best team out there, and that's how we train," Weckbacher adds. "We want to make sure if there are victims, live people, after an incident like that, we give them the best possibility of being found."

Since Sept. 11, NDSDF, the grassroots SAR organization housed in an anonymous terra cotta building in rural California, has received almost $1 million in donations. Dog lovers from across America donated to the foundation. Other donations came in from the families of victims. "One family lost their 26-year-old daughter," says Lori Mohr, an NDSDF training board member and volunteer. "They sent in a bunch of checks, saying that in lieu of flowers they'd asked people to give a donation to the foundation." A New York firefighter who lost his best friend sponsored a search and rescue dog through training. These donations have allowed NDSDF to put 18 dogs into training and establish NDSDF training groups in Florida and Ohio. While FEMA's goal of 300 certified teams is a long way off—the number hovers around 114—NDSDF continues to work toward that goal.

This fall, an American Kennel Club–sponsored art exhibit of hand-painted, life-sized fiberglass dogs will be displayed in public places around New York City to raise awareness of the need for more trained SAR dogs. In November, the statues will go on auction. New proceeds will go to SAR groups, including NDSDF.

Chat Room

1. What audience is the dog rescue feature targeting? How can you tell?
2. Where else could this feature be published? Why?
3. Mehus-Roe's story runs long for magazine publication. If it were to appear in a newspaper such as USA Today, how could it be shortened for readers who expect shorter stories?
4. If a reader encounters this story 10 years from now, will the reader be interested? Why? Why not? How might the topic be adapted 10 years from now?

Create

1. Write a query letter pitching Mehus-Roe's feature to a general audience rather than a niche publication.
2. Research other dog rescue teams and compare them to the SAR teams in your feature story.
3. Rewrite her story for publication in USA Today.

View

The magazine industry experienced a boom in the last decades of the 20th century but has suffered recent advertising revenue and subscription declines. This makes breaking into the industry difficult. Nevertheless, niche audiences assure that magazines are everywhere. To break into magazine writing, writers need to send engaging query letters to exactly the right publication. Magazine writing is linear and generally longer than newspaper writing. The time from story assignment to publication will be considerably longer. Magazines also tend to use better-quality, larger-size illustrations in greater numbers than newspapers do. Though feature writing is still feature writing, those who write for magazine face tasks and responsibilities unique to that medium.

Help:

cut line   caption that appears underneath a printed picture to identify who appears in it and to explain the action or context.

niche magazines   publications that target specific audiences and interests. Niche magazines have grown during the past two decades.

query letter   a brief, ideally one-page, letter that "hooks" an editor on a story idea, details how the story will be researched and developed, and mentions available photos and illustrations, expected length and delivery date.

Writer's Market   a reference book that provides an invaluable resource for freelance writers, offering information on wide publication market categories as well as advice on a variety of issues of concern to freelancers.

Notes

1. Amelia Michalski, a writer from Morris, Ill., contributed this reflection on the magazine.
2. Fourth quarter year-to-date advertising lineage figures compiled by Elizabeth Sturdivant put *American Way* in the category's top position in *Advertising Age* 4 Feb. 2002:14.
3. "Data Center," *Advertising Age* 29 July 2002:18.
4. Mercedes M. Cardona and Bradley Johnson, "The Ad Market," *Advertising Age* 12 Jan. 2004:8.
5. "Special Report: Magazine Forecast," *Advertising Age* 22 Oct. 2001:S-1.
6. Jon Fine, "Magazine World Weighs Its Worth," *Advertising Age* 22 Oct. 2001:S-2.
7. Monle Lee and Carla Johnson, *Principles of Advertising: A Global Perspective* (New York: The Haworth Press, 1999) 192.
8. Jon Fine, "Tongue Tied to Brand," *Advertising Age* 17 June 2002:14.
9. "Special Report" S-12.
10. "Special Report" S-12.
11. "Special Report" S-6.
12. "Weighs Its Worth" S-2.
13. "Special Report" S-12.
14. Kathryn Struckel Brogan, ed., *Writer's Market* (Cincinnati: Writer's Digest Books, 2003) 49–51.
15. Brogan 8–9.
16. Brogan 70.
17. Brogan 16.

8

The Newspaper Industry

There is something intrinsically political—and strongly democratic—about literary journalism, something pluralistic, pro-individual, anti-cant, and anti-elite.
    —*Mark Kramer,* Literary Journalism[1]

I climbed the grand marble staircase of The Tribune Building in June 1974, a skinny college student in a blue, bell-bottomed jumpsuit. As will be the case for many of you, my break into newspaper came through an internship. That wondrous summer I earned three journalism credits, flew with a stunt pilot, walked on building rooftops and participated in a master dance class. My theater and dance performance history brought assignments for arts features and reviews; I expanded my writing portfolio from tearsheets of stories I wrote for the college newspaper to tearsheets from the real thing.

It was the beginning of a beautiful friendship.

Almost 30 years later, I am still climbing those same marble stairs, now for site visits for my student interns. The people in the newsroom at the top of the stairs still sit in the gray light of their computer terminals, fingers tapping out stories at no-frills metal desks. The days of visored, cigar-chomping editors may be gone, but the "uniform" has changed little—tailored slacks and striped shirts, an occasional blazer. Nothing glamorous. The newsroom remains as dark and cavernous as ever. In the dark behind the stairs, presses continue to roll out the daily.

A decade ago it was predicted that the Internet would sound the death knell for the newspaper industry; "newspapers continue to struggle with readership declines, particularly among young people."[2] Nevertheless, newspaper ad spending outpaced magazines in the first half of 2002 with a 6.3 percent increase (compared to magazines' 3.7 percent drop). By October 2003, newspaper ad spending increased 15.2 percent over the previous year, showing an impressive post–Sept. 11 recovery.[3]

While the most important distinguishing feature of the magazine is its ability to pinpoint a specific reader demographic, the strength of the newspaper is its ability to provide complete coverage, unrestricted to any socioeconomic or demographic group. Mark Kramer's description of literary journalism could just as well describe newspapers. Almost everybody reads them.

The Newspaper

From tabloids to newspaper chains to Woodward and Bernstein to the USA Today revolution, newspapers in the 20th century underwent dramatic change. When Al Neuharth launched USA Today on Sept. 15, 1982, "the first national newspaper distributed by satellite to local printing plants,"[4] the "innovative, colorful and breezy" format transformed the industry.[5] Though initially criticized as "shallow," the national newspaper grew to boast the largest U.S. circulation (See Table 8-1). Tom Curley, the newspaper's current publisher-president, may speak for the entire industry when he says, "we are always evolving." Curley views the "third-generation" USA Today as a "network. We feed content to television. We feed content to the Internet from the same core platform."

Of course, it has always been newspapers first. Other media have long followed the newspaper industry's lead.[6] When I interned in a television newsroom, I watched

TABLE 8-1   *Top U.S. Newspapers (based on paid daily circulation)*

1. USA Today	2,155,000	Gannett Co. Inc.
2. The Wall Street Journal	2,091,000	Dow Jones
3. The New York Times	1,119,000	New York Times
4. Los Angeles Times	955,000	Tribune Co.
5. Washington Post	733,000	Washington Post
6. New York Daily News	729,000	
7. New York Post	652,000	
8. Chicago Tribune	614,000	Tribune Co.
9. Newsday	580,000	Tribune Co.
10. Houston Chronicle	553,000	

Adapted from "Data Center," *Advertising Age* 10 Nov. 2003:11, and 28 Oct. 2002 <www.freep.com/jobspage/links/top100.htm>. The Tribune Co. boasts three newspapers in the top ten, while the Gannett corporation's USA Today remains the nation's number 1 newspaper. Except for the Washington Post USA Today and the Los Angeles Times, many of the top papers saw modest circulation increases in spite of the recession. Among the top 10 from 2002 to 2003, only the Chicago Tribune experienced no change in weekday circulation averages. The Dallas Morning News remains out of the top 10, a place it occupied in 2001. Two Chicago papers made the top 25 with the Chicago Sun-Times at 13.

the news director and his staff begin the day by reading and quoting from the morning newspapers. Although for years newspaper "could not beat television and radio with breaking news," now the online presence of every major newspaper drives the media agenda.

Although newspapers are characterized by wide coverage, they still have the ability to target geographically. The South Bend Tribune serves a wide market area; published in South Bend, Ind., the newspaper produces special editions for its Michigan readers and for the smaller Indiana towns. Some newspapers publish special editions for neighborhoods and suburbs.

In addition to their wide coverage and broad audiences, newspapers are also distinguished by their timeliness. I'm sure you've never heard of a daily print magazine—such an animal simply doesn't exist. Short lead times allow newspapers to get news out quickly, even to make last-minute changes as news events emerge. Newspapers are classified as daily, weekly, national, special audience, organizational (in-house newspapers for employees and other internal publics), ethnic press and penny saver (local newspapers heavily subsidized by advertising and delivered free of charge to all addresses in a geographic area).

With the unprecedented increase of ethnic populations (especially Hispanic) in the United States, the ethnic press category has gained importance. New York City's Independent Press Association now runs a nonprofit ad-buying agency to help advertisers reach up to 3 million ethnic New Yorkers through over 100 immigrant, ethnic and community newspapers.

Like magazines, newspapers appeal to those interested in reading, so editors try to publish, whenever possible, the longer, more reflective feature story. Special sections and editions give space to in-depth stories that are not necessarily cutting edge. However, unlike magazines, newspapers tend to be more quickly thrown away or recycled.

Newspapers were traditionally printed by the rotogravure process, geared toward speed rather than quality. Modern offset presses produce higher-quality color reproduction, but readers still complain that the ink comes off on their hands. Newsprint paper is inexpensive and its quality inferior to magazine's heavier, glossy paper. Newspapers typically come in two sizes: tabloid (14" × 11") and broadsheet (22" × 13", divided into six columns). Newspaper size invites visual clutter with multiple headlines, stories and a myriad of advertisements cutting across the page. Producing reader-friendly layout can be quite a challenge, but the advent of Quark Xpress for page layout has made the editor's job easier than it was with the old-fashioned cut-and-paste method.

The Newspaper Feature

Typically, magazine and newspaper features differ in length and in the time frame you have to write them. Of course, exceptions to the length differences exist—Sunday newspaper magazines publish two- to six-page feature spreads, and magazines publish feature briefs and columns. Nevertheless, magazine "pieces traditionally are much longer length than newspaper pieces" and allow more than "a small amount of space" for background.[7] A newspaper might get around length restrictions by running a narrative series, as The South Bend Tribune did with the River Park rape story (see Chapter 4).

The same difference-by-degree applies to the time frame from idea to publication. The time cycle for a feature story to be published in the national or local news section of a newspaper could be as short as an hour; it could be years after the query has been written before a magazine feature actually appears in print. However, editors of newspaper feature sections generally allow the writer a week or even a month to work on the story.

Magazines and newspapers also differ, by degree, regarding who will write the story. Magazines keep a limited staff of full-time writers; associate writers may also write for other publications or have full-time employment elsewhere. That's the reason the freelancer's query was covered in the previous chapter.

On the other hand, newspapers rely primarily on their staff writers. An old rule of thumb was that a newspaper should have one full-time writer for each 1,000 of its circulation.[8] Layoffs caused by the current recession have increased the workload for staff writers and led to greater dependence on stories from other sources—wire services, *stringers* or special assignment writers, correspondents and public relations releases.

Types of newspaper reporters include *general assignment writers, beat reporters* and *special assignment writers*. As the name implies, the first type may be assigned to any section or story. Beat reporters cover specific areas, such as the police or business beat, or may be assigned to specific sections (I know someone who got her start in newspaper by writing the obituaries). For decades, I worked as a special assignment writer for The South Bend Tribune's arts and entertainment sections. I liked it because I could suggest, accept or reject assignments. Even though it wasn't a full-time position, my importance to the paper was recognized when I became the *lead reviewer,* the writer who covered the major arts events.

Depending on the assignment, the newspaper feature writer may write as many feature types as there are magazine categories: lifestyle, automotive, business, sports, food, bridal, opinion and advice columns, travel, television, technology, health and so on.

Newspaper writers "must be able to work on more than one story at a time."[9] If you can't chew gum, talk on the phone and do your homework at the same time, you probably won't want to do newspaper work.

Like magazine writers, newspaper feature writers look for story ideas related to "life and life's issues"[10] and cover feature topics from a variety of angles through observation and multiple interviews. More than magazine writers, though, newspaper writers need to have the coping skills to come back to the office and write a story even when an interview was denied or went flat. "Always have Plan B," advises Eric Hansen, sports editor for The South Bend Tribune.

Newspaper writers have traditionally used an objective tone and employed third-person, active voice, but today's style is moving toward the conversational.

Depending on the section of the newspaper to which the writer is assigned, this print publication offers a variety of feature writing opportunities. Some of the traditional features types are described here.

Historical Features

Many local and regional newspapers give weekly coverage to area history; others print an occasional historical piece in feature or magazine sections. Student writers may overlook a wealth of campus history and legends. Professional writers love historical features, and editors find space to place them. The South Bend Tribune on Oct. 27, 2002, devoted space in each of its sections to Halloween lore. A newspaper may not carry a theme as completely as a magazine can, but holidays and news events of national or local magnitude may generate a theme that threads through the newspaper. The lifestyle section for The Tribune's "Halloween issue" featured three stories on local hauntings and a book review, all spanned by a single headline: "Spine Tinglers."

Inside the section, The Tribune picked up a Dave Barry column headlined "It's Scary, It's Creepy ... It's the Web Site on Halloween," that retold the history of Halloween customs, such as pumpkin carving, that Barry found on the Internet. A book review picked up the Halloween history motif: "Death Makes a Holiday: A Cultural History of Halloween" by David J. Skal. Finally, the weekly local history column, "Through the Years," told of a haunted flour mill that drew crowds in the 1840s.

Carl H. Giles advises writers of historical features to "parade" the history before the reader, "take him back to when it occurred. Color and description are essential. You must rebuild something that has been gone a long time, perhaps for centuries."[11]

The feature on Marz Sweet Shop that follows in this chapter appeared in The South Bend Tribune's Sunday magazine. It's longer than a regular-section feature and creates a hybrid type, the historical/business trend feature.

Business and Trends

Newspaper business sections publish news stories, business briefs and features that profile businesses and the people who work for them. A business feature might capture the unique (one of my favorites profiled a dog-poop cleanup service), the nostalgic (the Sweet Shop story in this chapter) or a new product, company or service (chair massage at the local airport).

A reader favorite is the trend feature. You know about trends ... what is "trendy," "the rage," the current fad. The Sweet Shop story, about a past trend in which teenagers hung out at soda fountains, explains why teenagers liked to do this, and explains why it is no longer a trend (malls and fast food restaurants). At the same time, the story conveys the history of a small, family-owned business that has existed for half a century. And it profiles the business with details typical of business features—its location, hours of operation, product lines, processes. Business writers face the challenge to flesh out their stories with human interest. Interviews with owners, workers and customers put humans in the story. Description puts the reader there.

Click Here: Checks *Facts* or *Fail*

If you don't have a passion for fact checking, which includes spelling names right, then you may not belong in journalism. Just ask the good folks at the Medill School of Journalism. Brandon Stahl's article in the Spring 2003 issue of Medill Magazine elaborates on "The Medill F." In one of the world's top journalism programs, Stahl reports, "a factual mistake, including the misspelling of a name," carries an F, "no matter how much you cry."[14]

Stahl documents the case of a student who called a street an avenue and received an F for her error.

The reason for this obsession with small detail? Medill professors give the Fs as a "warning that factual inaccuracies and misspellings in the students' stories, while bad at school, are even worse at work," Stahl explains.

Major magazines employ fact checkers; section and copy editors at newspapers serve as a last bastion against the dreaded factual error. In the end, though, it's the writer's responsibility.

The essence of journalism is that it is fact-based, not fictional, writing. If you don't understand the concern about whether or not there should be an "e" in Kelly, then either you don't belong in this profession or you just have never received the big fat Medill F from a fed-up professor. That experience apparently creates a passion for accuracy where none may have existed before. Stahl reports that Washington Post columnist Mike Wilbon, once a recipient of the Medill F, still wakes up at 3 a.m. from nightmares about factual errors.

Color Features

Color features sidebar news stories and help explain a news event "by providing atmosphere or mood—happy or sad, warm or cold, exciting or dull—for those who could not be there."[12] Because each writer "will gain a different perspective on the same event,"[13] the color feature reflects the writer's individuality. Holly M. James' Vidalia story (see Chapter 6) combines qualities of both the travel and color feature.

Long or short, color features often accompany news stories for background or "color," earning them a place in the repertoire of a newspaper writer. This type of

Sidebar: Resources for Newspaper Writers

The American Society of Newspaper Editors' Web site, http://asne.org, provides internship postings, job fair schedules and job listings as well as tips on starting a journalism career and landing a first job. The site links to conferences and other journalism groups.

story fully employs details, attitude and tone. The short color features in this chapter resulted from a feature writing class assignment. After watching the Eiffel Tower scene from the spy thriller "Company Business," the students were challenged to write color features to go with a hypothetical news story about a spy swap at the top of the Tower. Notice the opposite moods evoked in the two published here. Color features leave room for personality.

Sports

Writing a sports feature is like writing any other feature. "What counts is people, the personalities, not the stats or technicalities," says sports editor Eric Hansen. Technical jargon won't evoke emotions or engage readers; talk to them in language they will understand. "Everyone knows what it is to be booed," says Hansen.

Sports events carry statistics that may need to be reported; don't clutter your feature with them. You might include crucial stats in the feature itself, but consider reporting most of them in an *info box*, a *breakout* of no more than 12 column inches.

My impression had always been that action verbs mattered more with sports than any other kind of writing. Wrong! Action verbs matter in any kind of writing—journalistic or academic or imaginative. Whatever constitutes great writing works for sports, including accuracy. Hansen recalls a professor of his at Ohio State University who gave a zero on papers with wrong facts or misspelled names.

Finally, research backbones the sports feature. Hansen reminds writers to "research before you talk to someone. Sports figures expect you to know a lot of basic information about them. Instead of asking, 'Where'd you go to high school?,' talk to the high school coach before the interview. Then ask about something the coach told you."

Read the best sports writers. For example, feature writers for top publications such as Sports Illustrated spend a couple of days with people they profile and talk to family members and coaches.

Lifestyle

Newspaper lifestyle sections focus on homemaking topics such as bridal, food, crafts, fashion, and house and garden. These sections include regular columns (such as horoscopes, geneology, gossip, homemaking tips), news items (engagement notices, wedding announcements, new product announcements) and full-length features (seasonal food features often include recipes, and seasonal gardening features may warn about insect invasions or describe a unique local garden). Food features may include historical features (how the tomato, once considered poisonous, became a nutritional staple) or trend features.

Molly Strzelecki's food trend feature, "A Slice of Comfort," suggests that in the United States and Europe, post–Sept. 11 stresses have increased sales of "comfort food" such as good old American pie. Grab a snack and read the excerpt of her feature at the end of this chapter.

"Free Falling"

This color feature and "Dining High," which follows it, were written as student responses to the following assignment: First, they watched "Company Business," a film in which Gene Hackman and Mikhail Baryshnikov, portraying spies from the CIA and KGB, engineer a spy swap at the top of the Eiffel Tower; second, the students were to write color features that might sidebar a news story reporting the incident (on the hypothesis that the scene in the movie was real). Denise McGuire and Katie Miller wrote "Free Falling," and Christina Reitano and Holly M. James wrote "Dining High."

Of the six million people who visit the Eiffel Tower each year and take the elevator up to the top, some visitors have a unique way to reach the ground again. In 1984, two Englishmen parachuted off the top. In 1923 a journalist took a scary ride down the tower on a bicycle. Around 400 others took a different route, jumping or falling from the top of the tower. Unfortunately, they did not reach the ground since the tower widens from top to bottom. Their fragmented bodies had to be removed by local firefighters. When you are in the Eiffel Tower in Paris, please take the elevator.

"Dining High"

Don't feel like lugging that picnic basket full of goodies for you and your sweetheart to the top of the Eiffel Tower? Never fear, many tables await your arrival high above Paris. Altitude 95 and Le Jules Verne Restaurant sit high above the city and offer not only exquisite dining but also a breathtaking view.

Altitude 95, so named because it sits 95 meters above sea level, seats 200 and offers a variety of drinks and refreshments. It faces the Seine and Trocadero rivers; that's why it's modeled after an airship anchored over Paris.

Le Jules Verne offers an upscale dining experience 125 meters above ground on the tower's first level. It continues to be one of the top-rated restaurants in France. In the film "Company Business," the spies portrayed by Gene Hackman and Mikhail Baryshnikov choose to wait out a swat team of Paris police with some fine dining in Le Jules Verne. Baryshnikov's character doesn't even seem to mind that he was shot prior to his impromptu meal.

"The Newest Member of the Footwear Family"

When Jennika Kirkbride was assigned to write a trend feature in her journalism class, she had an idea that just seemed, well, flimsy, maybe even a little low compared to the high-browed ideas proposed by her professor. How could fashion socks rival such topics as trends in religious education or multigenerational relationships? As Jennika discovered in a conference with her professor, not all trend features need to focus on serious or provocative issues. Jennika wanted to keep her readers on their feet, or at least thinking about them.

People are always looking for a new way to express themselves. They dye their hair, wear outrageous clothing or pierce nearly any part of their body. But sometimes these unconventional options are just not for everyone.

Instead, many people are picking up on the new way of self-expression through what they wear on their feet. The plain white cotton socks of yesterday are being replaced on shelves and in shoes by colorful fashion socks that do everything from express moods to celebrate birthdays.

Jill Thomas of Dillard's department store in Toledo, Ohio, indicates that sales of fashion socks have been on the rise in the past year and "especially during this past Christmas season; they make great gifts." Generally priced between $5 and $10 per pair, they make quick and easy presents that Thomas says will please anyone.

Kristina Kitzler, a sophomore at the University of Toledo, attempts to explain the attraction. "White gets boring day after day and sometimes you just need to switch it up," she says as she points to her feet and her Hawaiian-themed socks, a gift from a close friend. "These kind of commemorate a movie we love." Kitzler brings up the two main reasons the new breed of socks has become so popular: variety and commemoration.

Jule Brown of Kohl's department store explains, "People use them to break up the fashion monotony they have come to expect from a pair of socks." The colorful designs encompass a wide range of patterns, including solids, stripes, words, holidays and special occasions, and an extensive variety of themes. Brown holds up a light blue pair covered in fluffy cumulus clouds. "This is one of our most popular styles, very versatile whether you're enjoying a sunny summer day or wishing for one in the middle of winter."

While it's true that holiday and special occasion socks have a very short life span, seemingly appropriate only one day a year, this doesn't deter people from picking them off the shelves. In January alone, customers purchased over 300 pairs of fashion socks from one department store.

The need for a change of pace is not the only factor behind these numbers; as Kitzler points out, there are reasons individuals pick a certain pair of socks. How do they choose the design that is right for them? Celebration, commemoration, remembrance and just plain attraction are some of the overwhelming inspirations.

The Web site http://JoyofSocks.com exhibits an extensive collection of novelty and fashion socks. There are socks to celebrate St. Patrick's Day, Hanukkah, the Fourth of July and every other holiday imaginable. There are designs to commemorate special events such as weddings, births and even the 21st birthday. They can commemorate an experience or a place, the beach, the desert or the city. They can also be chosen solely on personal attraction, a favorite color or an alluring design.

Stores may also offer a selection of higher-quality socks that bring a higher price based on materials and designer. Some of the higher-priced socks are made primarily of silk or cashmere. These generally price between $10 and $20, but some go as high as $45 a pair.

Women are not the only ones hanging up their predictable, conservative socks. Men everywhere are slipping new subtle designs into their wingtips before heading to the office. Brown says, "Men are also sporting the socks. They, of course, have to retain a little more conservatism than women and girls do, but we do have something for them, too."

The newest member of the footwear family is catching the attention of people in all categories. They can be found in most department stores, specialty stores and Web malls. While they have yet to take over the versatility of their predecessors, the new socks are here to stay. So before you lace up, pull on or tie tight, remember to let your shoes do the walking, but let your socks do the talking.

FEATURE STORY

"Marz Sweet Shop"

Carla Johnson
Author of "Marz Sweet Shop" as well as "Troy Donahue" (Chapter 2), "Colm Feore" (Chapter 3), "Roxie Hart" (Chapter 4), "James Earl Jones" and "Tom Wopat" (Chapter 5), "Marin Mazzie" and " 'Rent' Reclaims Broadway for U.S." (Chapter 6) and two reviews (Chapter 11). Photo courtesy of J. Wolfe Photography.

This hybrid story (a combination of business, historical and trend) appeared in a hybrid publication (a newspaper's Sunday magazine supplement). Published in Michiana Magazine on Sept. 22, 1985, the story caused an influx of visitors from as far as 200 miles away from this little Michigan town. We maintain the story breaks of the original publication (the magazine breaks long stories visually to enhance readability and to defray a copy-heavy impression). The story has undergone some revision, however, just as Marz Sweet Shop has. Kathleen Marazita, now a widow, runs the shop alone now. On weekdays, she turns the CLOSED sign over and turns on the lights in the front half of the store. Yet the candy apples and hand-dipped chocolates still taste the same.

ON FALL DAYS, at lunchtime or after school, students kicked through the leaves on the Front Street hill heading toward downtown Buchanan, Mich.

It was back in the days when the girls from Buchanan High School bought their Tangee lipstick and seamed stockings at the five-and-ten, when Jim Jackson would stop at home after school just long enough to yell, "I'm going to the Sweet Shop, Mom!"

Jackson is now retired from 33 years as a middle school teacher, but he still stops in at the Sweet Shop occasionally to buy homemade dark chocolates. Half a century after Jackson and his buddies piled into its vinyl booths, the Marz Sweet Shop opens a few hours a day, with lights in just the front half of the store. Malls and fast-food restaurants have taken their toll on the small, family-owned soda fountain and candy store.

For the Buchanan youth of yesteryear, the Marz Sweet Shop at 205 E. Front was the official hangout. Within its maize-colored walls, youngsters sipped flavored phosphates or enjoyed a Buchanan High School sundae, and put their change, which weekly totaled into the hundreds of dollars, into the jukebox at the back of the room. Students knew the rumors—the jukebox was serviced by the local Mafia—but no one cared.

What they played was early Elvis, Sam Cooke, Ricky Nelson. Throughout the '50s and '60s, Marz Sweet Shop accommodated some 500 customers each day.

For those who entered under the red and white candy-striped awning, the Sweet Shop was a place to meet friends, make new ones and cultivate romance. "People got engaged here," Paul Marazita, the shop's owner, recalls.

Marazita bought the building for the shop in 1946 and spent a year converting an old variety store into a modern structure. "To go into the store in 1946 was to walk into the 19th century," Marazita says.

He completely remodeled, changing the "old-time front" into a facade with a modern look, lowering both the ceiling and the floor inside. He and his partner-wife, Kathleen, did not consider renovation. "At the time, modern was the way to go," she says.

Marazita had just come home from the Army, taken a job as a policeman in Niles, Mich., and married Kathleen Stoner, whose parents owned a large farm north of Buchanan. But the "sweet" business was in his blood.

His father, Frank Marazita, had come from Italy some years earlier and founded the Michigan Fruit Co. in Benton Harbor, Mich. Frank Marazita then started a confectionary store in Niles with his brother. Paul Marazita's brother, Albert, still runs the Veni Sweet Shop in Niles.

PAUL MARAZITA says he has worked in a candy store for "probably 50 years of my life, from childhood."

He opened his own doors in Buchanan in 1947, featuring fountain service, light lunches and homemade candy. He worked 16 to 18 hours a day, seven days a week, to keep up with making ice cream, candy, phosphate flavors and sundae toppings from scratch. He opened the shop at 8 a.m. and wouldn't roll into bed until 1 or 2 in the morning.

For a time, he and Kathleen lived in the apartment above the store, but the beginning of their family of six children necessitated that they move out into a house.

"I love the business," Paul Marazita says. "You've got to like this business to be in it. 1957 was the last vacation I took."

He said that long hours were required to keep up with competition. "Everybody else was open till 11 at night" in downtown Buchanan, he recalls, "so I had to stay open, too. The only holidays we closed were Christmas and Easter."

Buchanan had five or six fountains open in those days, but "they didn't all have complete fountains. The drugstore fountain was just a leader item to bring people in."

Marazita attributes the prominence of his shop as a teen hangout to several factors. "We had beautiful girls working here," he remembers, "several Miss Buchanans, in fact. They worked on a co-op basis through the high school to help cover the lunchtime business. That helped a great deal."

With six children of his own, Marazita became involved in a number of youth activities and knew many of the youngsters through this participation. Essentially, he just liked kids.

"I KNEW the kids' problems and tried to help solve them," he says. He remembers one young lady who came into the shop "with her little bag and everything, running away from home." Marazita talked her into staying for a soda, then went around the corner to where her father worked at the bank.

Her father came back to the shop with Marazita, the problem was straightened out, and a runaway was averted.

"I had a free phone here," Marazita says. "Kids could come in here after school or the movies or the football games, call their parents and have them pick them up. Their parents didn't worry about them. They knew they were either at the movies or here."

Marazita also ran a tight ship. There was no nonsense. He remembers asking a young man to leave because he had put ice down a girl's back. Sometimes he would require an unruly youngster to sit at the counter until he cooled down. "Those kids never even left a napkin on the floor," he says.

As a testimony to Marazita's importance to the youth of the '50s and '60s, he was asked to be a featured speaker for a 1959 Buchanan High School class reunion.

Paul Marazita carefully selected Buchanan as the location for his store. Studying census records, he noted the town's rapid growth during the 1940s because of the growth of Clark Equipment Co. He figured if the town kept growing at that rate, it would soon reach a population of 10,000. But that didn't happen.

Times change, and so has the clientele visiting Marz Sweet Shop. It wasn't the departure of Clark Equipment that hurt Marazita's business though. The number of students who visited the store started to decline when they were kept at school during the lunch hour. Today just a handful of local youths venture in, but the ghosts of the old crowds linger on.

Marazita says old-time customers are always dropping in. "When they come back to Buchanan from out of town, they always bring their children in," he says. That gives him the opportunity to make good on his promise to treat their children to their first malted milk.

DOWNTOWN Buchanan is also not what it used to be. It was once an ideal location for such a business. "The movie house and the grocery stores used to be downtown," Marazita says. His current business is just "20 percent of what it used to be."

In some ways, the decline has not been all bad. "I didn't know the business would be so confining," Kathleen says of the early days. The store now opens for shorter hours, and Kathleen can operate the shop with little or no help.

The business has become seasonal. Ice cream is the big summertime seller, and candy remains the mainstay, especially around holiday time.

"We have chocolate Santas at Christmastime, Easter rabbits and baskets, and caramel apples September through October," Paul says. He has offered more than 100 varieties of hand-dipped chocolates, designed by Kathleen. For Valentine's Day, customers can choose from heart-shaped boxes varying from a half-pound to a gigantic 10-pound box which measures 26-1/2 by 26-1/2 inches. Customers can choose which candies they want packed into the boxes, making up their own assortment.

Though the young trade has diminished, the store now serves more adults. The jukebox remains, but it now survives longer. "I used to get a new jukebox every six months," he recalls. "The kids would wear it right out."

Marazita still makes his own candy, ice cream, phosphate flavors, sandwich fillings and sundae toppings. "I just made 30 pounds of candy this morning," he said on the day of the interview.

Though he makes divinity only in the summer, a dizzying number of other varieties is available year round. Among the more unusual selections is the "Marz special," crushed nuts and chocolate housing a maple cream center, and "chop suey," a blend of crushed nuts, raisins, coconut, and dates in a chocolate base.

MARAZITA USES pure chocolate with no emulsifiers, preservatives or fillers. He buys No. 1 peanuts that he roasts himself.

The candy recipes have been, for the most part, handed down from his family. Many of them were brought by his father from Italy. He says that they are all written down, even though he now carries most of them in his head, but "are not for publication. Candy makers are very tight about their recipes." He also stresses that part of the art of candy making is in the technique as much as the recipe.

"You have to be meticulous about each step," he says. "Miss a degree on the candy thermometer and the whole batch goes right out in the back. If someone comes in and says, "How was the Notre Dame football game, Paul?" a whole batch can be ruined.

"Candy making is an art. When you make candy by hand, it's very different from candy made by machine. But today there are very few candy makers. It's a lost art."

Every town, he says, used to have a confectionary store. He remembers going to a store in Grand Rapids, Mich., in the old days to learn to make taffy from a candy maker there. "I stayed there the whole day and watched," he says.

He especially remembers a candy maker he met in Chicago. "She dipped with both hands," he says with visible admiration. "She dipped candy like that all day."

Humidity, he says, affects candy making considerably. "It's a whole different ball game when you make candy in the summertime." Starting with Labor Day, he offers caramel apples, made from homemade caramel and fresh local apples.

Marazita also learned the art of making ice cream. He bought a five-gallon stainless steel ice cream maker and learned how to use it from the man who sold it to him. "I watched and wrote it down," he remembers.

HE MAKES his ice cream from a 12 percent butterfat mix. The legal minimum for ice cream in Indiana is 7 percent, he says. He features more flavors today than he did when he opened in 1947. Among the flavors he offers is tutti frutti, Italian for "all fruits," with cherries, pineapple, strawberries and gooseberries.

Other menu items include milk drinks (shakes and malts), 12 different kinds of sodas, and 12 drink flavors, in addition to any flavor phosphate. Plain and fancy sundaes range from $1.15 to $2.50 (in 1947 they were priced from 15 to 30 cents). A giant sundae feeds two people. Some have unusual names: Football, Henry Ford, Leave It To Me, Polar Bear and Sweet 16.

"EVERYTHING IS SERVED in glass, not paper," he says, "and we still make an old-fashioned soda." Marz Sweet Shop still features a complete soda fountain—it's one of the few places where customers can still enjoy phosphates.

The Sweet Shop is now open from 10 a.m. to 5:30 p.m. daily but closes at 5 p.m. on Saturday and all day Sunday. "We used to get a lot of people from South Bend in on Sunday, but we don't anymore." He still gets occasional customers from Chicago, Detroit and Milwaukee.

"People from larger cities really appreciate the store," Kathleen says. She recalls a couple of artists from Milwaukee who came in recently and took endless photographs. Their interest may have been motivated by the fact that the shop's interior resembles a pop-art diner.

Paul Marazita and his brother considered renting mall space for a new store but found the rent prohibitive. The only definite plan for the future is to continue in the business. For a man who once wrote his high school senior thesis on his future plans to own his own business, his life's destiny seems fixed. And though he finds himself still immersed in an old family tradition, none of his children seems likely to follow in his footsteps.

Dim fluorescent lights in the fountain area at the front of the store make the maroon booths in the back look dark. The jukebox lights flicker, but no music plays.

"I encouraged my kids to go to college and enter other professions," he says.

Chat Room

1. Would it be possible to write for a newspaper and magazine simultaneously? What would be the challenges of doing this?
2. Discuss differences between the newspaper and magazine features in Chapters 7 and 8.
3. In a group, discuss your personal areas of expertise, such as a business minor, participation in a competitive sport, and so on. What section of the newspaper would best utilize this special experience or knowledge?
4. "Roxie Hart" in Chapter 4 is a full-length color feature. Compare/contrast the full-length color feature to the briefs printed in this chapter.

Create

1. Write a news story about a current trend. Rewrite it as a feature.
2. Jennika Kirkbride wrote about socks; in the following story, Molly Strzelecki writes about pie. Come up with a lifestyle topic of your own and write the story.
3. What daily newspaper section do you read most often? Write a feature for that section.

FOOD FEATURE

"A Slice of Comfort"

As assistant editor, Molly V. Strzelecki writes columns and features for the trade magazine Snack Food & Wholesale Bakery. The tabloid page size "eats up copy," she says, so her food features tend to be longer than those published in newspapers. An excerpt from her five-page magazine story follows, cut down to newspaper length (an ellipsis denotes a cut). Sidebars listed the top five frozen pie brands and top five pie vendors, with Mrs. Smith's and Sara Lee topping both charts, and ads for Colborne's Pies Plus and Brolite baking products, related products not mentioned directly in the story, were placed next to the story on its jump pages.

In these trying times, America is not the only stressed-out nation in the world. According to just-food.com, a Web site that monitors a host of goods and consumer trends, half of Europe is stressed out as well. The factors involved really come from the same source. Stress rises when time-crunched consumers try to balance the demands of work, family and long commutes against an equally strong need for a semblance of a social life. In some cases, the stress is only "mild." In other cases, it can be "severe," especially if someone is also, for example, going through a midlife crisis.

While Europeans may seem to have different tastes compared to Americans, all share certain behavior quirks when it comes to relieving common stress factors.

We all eat.

"Stressed-out [Europeans] are proving to be an ad man's dream as they desire familiar, convenient brands and products and are willing to dig deep into their wallets in order to de-stress," says the online publication.

European comfort foods vary from country to country. What's considered comfortable for the British certainly may stress out the French and Italians.

For most Americans, despite our status as melting pot of the world, comfort foods are easy to identify. They're usually anything labeled with such words as "okay in moderation."

As for all-American comfort foods, what's more soothing than a slice of pie, any time of the day or year? While pie producers' biggest selling time is the traditional fall and winter holiday seasons, ongoing promotional campaigns and the onslaught of inventive desserts that rival restaurant quality are luring more consumers to the freezer case year round.

Heightened product quality and aggressive marketing, such as tying pie production to NASCAR racing, have helped to spark sales. Overall, the frozen pie category is up 9.6 percent, according to Chicago-based Information Resources, Inc.

"This year, it's probably not a surprise to most people that we've seen a huge interest in comfort foods, with consumers desiring to get back to basics and basic old traditional favorites," says Polly Johnson, vice president of foodservice marketing for Edwards Fine Foods, Norcross, Ga. "Products like your classic deep dish homemade apple pies, cobblers, crisps and other types of rustic, comfort desserts have been very popular this year."

. . . By adding such twists as caramel or chocolate toppings, exotic fruits and spices, pie producers have transformed pies into premium desserts.

. . . New or old, one thing reigns true. Pie is a quintessential North American icon that represents family and other traditional values. Mrs. Smith's Bakeries, with its new line of Soda Shoppe Cream Pies, is targeting time-pressed consumers who are nostalgically yearning for slower times and simpler days.

Working with the Haverton, Pa.-based DePersoco Creative Group, Mrs. Smith's developed a packaging design that's contemporary but contains overtones of old-fashioned goodness. The goal was to capitalize on the nostalgia of the 1950s, when small-town innocence thrived and families sat on their front porches waving to neighbors as they walked by.

Perhaps because of consumers' long-time love affair with pies, many producers remain optimistic about the category's future.

"Pie is one product that is an established business," says Leclerc, whose company entered the United States just a few years ago. "If people like it, they will always buy it, and that's good."

Chat Room

1. Lifestyle features, especially those about trends, almost always connect in some way to business and commerce. Do the features written by Jennika Kirkbride and Molly V. Strzelecki show similarities to a business feature? Defend your argument with specific examples.

2. Bring in lifestyle feature examples from a local or national newspaper. How many can you find that deal with weighty, serious topics? How many deal with light ones?

3. Bring in business features from a local or national newspaper. Compare story style and content to features found in the food or lifestyle section of the same paper.

Create

1. Research the history of a nutritional staple, such as bread or fruit (how about the orange, kiwi, pumpkin or corn?).
2. Write a food feature incorporating this history.
3. Look for food features on the Internet. Where do you find them?

View

Whereas the magazine industry tends to target a specific readership, newspapers have to reach a broad audience that spans all ages and demographics. Newspaper writers must write in a style that will appeal to a broad audience and also appeal to a wide range of needs and interests. For this reason, newspapers have numerous sections, many geared toward very different audiences and interests and written in different formats. Some of these formats include historical features, business and trend features, travel and color features. Readers toss out their newspapers without feeling much guilt, so stories must be timely and are unlikely to have the shelf life of a magazine feature. Readers want current information on a daily basis. Newspaper is a little like the fast food of print publishing—readers want stories that are quick and easy to digest. Of course, newspapers publish magazine sections and news magazines, the newspaper/magazine hybrids.

Notes

1. Norman Sims and Mark Kramer, eds., *Literary Journalism* (New York: Ballantine, 1995) 34.
2. Richard Linnett, "Magazines Pay Price of TV Recovery," *Advertising Age.* 2 Sept. 2002:28.
3. Mercedes M. Cardona and Bradley Johnson, "The Ad Market," *Advertising Age* 12 Jan. 2004:8.
4. George Rodman, *Making Sense of Media* (Boston: Allyn & Bacon, 2001) 83.
5. Ira Teinowitz, "USA Today: 'No Longer a Newspaper,' " *Advertising Age* 9 Sept. 2002:16.
6. Rodman 68–69.
7. Beverly J. Pitts et al., *The Process of Media Writing* (Boston: Allyn & Bacon, 1997) 298.
8. Pitts 303.
9. Pitts 300.
10. Pitts 297.
11. Carl H. Giles, *The Student Journalist and Feature Writing* (New York: Richard Rosen Press, 1969) 125.
12. Bruce Garrison, *Professional Feature Writing*, 3rd ed. (Mahwah: Lawrence Erlbaum, 1999) 141.
13. Garrison 144.
14. Brandon Stahl, "The Medill F," *Medill* Spring 2003:22.

Help

beat reporters newspaper staff writers who cover specific areas or sections, such as crime or business.

business feature a story that may capture a unique commercial experience, a nostalgic place or a new product, company or service.

color feature brief or full-length stories that sidebar news stories and help explain an event by providing atmosphere, mood or background information.

historical feature a story that seeks to re-create past events or profile historical persons to bring the past alive for present readers.

info box a sidebar that allows writers to avoid the inclusion of crucial stats in the feature itself. Also called a breakout, an info box generally runs no more than 12 column inches.

general assignment writers newspaper staff writers who may be assigned to any section or story.

lifestyle feature a story that focuses on home-related topics such as bridal, food, crafts, house and garden.

sports feature coverage of sports events that focuses on people and personalities more than just statistics, technicalities or technical jargon.

special assignment writers writers who are not employed on a newspaper's staff but regularly receive freelance assignments from the publication. They are also known as *stringers*.

9

The Internet

Art reacts to or reflects the culture it springs from.
—Sonia Sanchez, Black Women

Online entertainment features range from the audio ("Hear Jewel's Latest: It Sparkles") to the visual ("Scariest Rock Stars: See photos"). The New York Times on the Web lists a long menu of feature story categories that correspond to the newspaper's print sections. While the categories are the same, the presentation of the stories differs from print to Internet.

"Transitions like the one from print to electronic media do not take place without rippling or, more likely, *reweaving* the entire social and cultural web," award-winning essayist Sven Birkerts has argued.[1] The Internet has restructured "the reception and, in time, the expectation about how information is received."

A user who clicks on The New York Times' magazine category will find large photos, intriguing headlines, writers' bylines and *teasers*, short verbal messages designed to keep reader interest. The teaser is borrowed from television, a medium that encourages viewers to "stay tuned" for the next program and sometimes overlaps closing credits from one program with teasers about upcoming ones. Print magazines also employ *teasers* on their covers and contents pages to entice readers to buy the magazine or turn to major stories.

Of course, print and television do not have the interactive capabilities of the Internet. The online story typically provides interactive options, such as icons to click that lead to other, related stories, surveys or *photo galleries*.

On the Times Web site, users who click onto a <u>Go to article</u> icon will find another story, additional pictures and, presented as sidebars, more links to related articles and readers' opinions. The features on the Times Web site tend to run very long, sometimes covering over 30 screens with text. Apparently, the Times presents its online stories as *shovelware*, a story presented online with the same content as its print publication. The publisher lifts a print story from paper and reproduces it, as is, online.

Nevertheless, the user will not read the story the same way a print reader will. Birkerts says the Internet reading pace is rapid and story contents, unless printed out, become "evanescent. They can be changed or deleted with the stroke of a key."[2]

While print publications "exalt the word," the Internet "reduces it to a signal, a means to an end." Although we are still experimenting with the best way to write stories for this new medium, the content of most online stories may differ little from their print versions, but the way they are presented differs vastly. Directed to stories through traditional headlines and teasers, the print reader reaches the end of the journey with the end of the story. The online user may click on the story and never read it all; the story may simply link the user to other options.

The Internet Industry

Based on advertising impressions and number of visitors to its shopping site, e-commerce pioneer Amazon.com remains at the top of its field, with books the top product category among online shoppers.[3] This fact reveals something about the Internet user, backing up earlier research that revealed users to be "younger, more educated and more affluent than the country's general population."[4] They're people who *read*. Although naysayers in the 1990s worried that the Internet would spawn generations who wouldn't read and write, this is not turning out to be the case at all. An advertisement for The Wall Street Journal summarizes the attraction of online news sites: "Online—because it provides depth of content and the power of sight, sound and motion" as well as "iconic imagery" that can be recognized "around the world."

However, the peak in traffic to news sites "in the weeks immediately after the [Sept. 11] attacks" preceded a traffic decline that was expected to be permanent.[5] Despite this, Internet ad spending increased 22.3 percent between October 2002 and October 2003, compared to just a 6.7 percent ad spending increase for magazine and 15.2 percent for newspaper.[6] Online news sites remain contenders among the information industries (see Table 9-1 for the online national news sites with the largest number of daily visitors post–Sept. 11).

No other medium can package information quite like the Internet.

Internet Story Structure: The Package

Britney Spears comes alive on AOL Time Warner's AOL Music channel via an *online package* that includes "live chat, access to songs, interviews, concert-tour information, the artist's Pepsi-Cola ads, backstage diary entries, and more."[7] Web site "packages" utilize various platforms and links to create "the power of sight, sound and motion" touted by The Wall Street Journal ad.

TABLE 9-1 *Top Online National Print News Sites (based on daily visitor traffic)*

1. New York Times	nytimes.com	1,015,000
2. USA Today	usatoday.com	628,000
3. Washington Post	washingtonpost.com	593,000
4. Time magazine	time.com	259,000

Adapted from "Interactive," *Advertising Age* 29 Oct. 2001:34, and Jupiter Media Metrix. Figures are from the week ending 14 Oct. 2001.

Think of the online story as a *package* that contains different levels of information, such as headlines, teasers, abstracts and links to related sites. Packages may also include interactive elements such as discussion questions, FAQs (frequently asked questions and answers), games and quizzes. Headlines are especially important as a way to entice site visitors who are faced with an unprecedented number of reading choices on the Web.

Packages may be enriched with subheads, bulleted lists and graphic elements (although graphics that take too long to download should be avoided). Options at the end of stories may allow users to e-mail the article to others or to print out the story. While the story remains the focal point of the package, as it does in print layouts, writers and editors tend to spend considerable time creating these packages. User expectations about the story itself have already been shaped by the online presentation method. Writers are just beginning to learn how to respond to these expectations.

Linear versus Nonlinear Writing

According to Birkerts, "The order of print is linear, and is bound to logic by the imperatives of syntax."[8] Traditional *linear structure* moves the reader from the story's beginning to its end in sequential fashion. In most ways the opposite of print's linearity, the electronic order moves information and contents along a network. Online writing takes a *nonlinear* structure in which a story appears in self-contained "chunks" with embedded links or multimedia elements—text, photos, graphics, sound, video and animation. Since online readers tend to scan rather than read word for word, journalists are beginning to refine the nonlinear story. *Shovelware* may eventually become obsolete, and print stories may increasingly be reconstructed when they are republished online.

Story restructuring could blur certain distinguishing characteristics of the feature. Since research shows that less than 12 percent of online users read stories in their entirety,[9] some experts recommend that the inverted pyramid structure be used for all stories. As far as length, text should be about half the length acceptable for traditional newspapers and magazines because of the reduced readability of the *pixel*. While print features have the time and length to provide "depth, imagination, scope" in a time when the need for reflection has "never been more pressing,"[10] the online feature tends to be written and published more quickly.

Yet Internet packaging allows users to create their own narrative story by piecing together story fragments via links. This capability makes the user a participant in the creation of the story and revolutionizes the idea of story itself. Writers and editors provide random story fragments, and readers construct the story according to their own interests and needs. For example, a two-screen story about a Florida court decision in the Elian Gonzalez case linked to another story that backgrounded the tragedy that brought the young Cuban boy to Florida in the first place. The link provided a *flashback* to help the reader understand the context for the court decision. However, a given reader might only have sought information on the tragedy at sea and might have been uninterested in the court decision. The initial story and its links served the purpose of driving the user to other sites. These linked sites might include photo galleries and opinion polls, all part of the story *package*.

Writing the Internet Feature

Just when I thought I'd learned to anticipate everything a newspaper or magazine editor might do to a story, I entered into a whole new writer/editor relationship when I submitted my first feature for Internet publication. I submitted "City of Light, City of Love," a travel memoir, to the Educational Travel Review, an Internet publication of the American Council for International Study (ACIS). I hardly recognized "Love City," the version published online.

The publication's editor, Theodore S. Voelkel, explained that he always looks "for the shortest title possible, preferably a pun, or a cliché spun differently. 'Love City' works off Cliché City, Grab City, Putdown City, Skin City, etc., where City means 'a lot of.'" In addition to shortening the title, Voelkel cut what he calls "connective text" to make "points short and punchy to keep the reader's attention. That's even more important on the Internet than in print publication. In the 18th century, my print counterpart would have labored to pump up the frills and fringes—that was a leisured society in spades." Voelkel edits his copy to avoid shovelware.

In addition to "connective tissue" cuts, Voelkel rearranged the order of the narrative. He felt that early background information created a "back-and-forth" that "asks the reader to do too many things at once—absorb a fairly complicated series of events." For the Internet, he says, "the pace needs to pick up," and paragraphs should "zip by as fast as possible. Zip zip zip, real quick." Voelkel admits he "took liberties, a detached, scalpel-wielding, heartless clinician (the editor) jumping into the thing cold, trying to boil down the action to its simplest components."

Whether we work with a print or online editor, though, we can count on one absolute. All publishers have deadlines: "At the time, I was just pressed and worked as fast as I could to get the job done," Voelkel confessed in his explanation.

In writing for the Internet, also expect to learn a new vocabulary and to spell the terms correctly. The Associated Press Stylebook now includes an "Internet guidelines" to "commonly used Internet, computer and telecommunications terms."[11] Those who write for or about the Internet should refer to these pages for guidance. For example, many still write "website"; even though the AP preference is "Web site."

Other spellings may also surprise you; for example, "homepage" is incorrect AP style. Until you've mastered the terms, make it a habit to always look them up.

The AP Stylebook guide also includes valuable information for writers on how to search for, use and cite online sources in a media story.

The Roundup: An Experiment

The New York Times' online newspaper includes a regular travel feature category. For a post–Sept. 11 issue, the site published feature briefs with one-word headlines indicating the story's destination, such as "Rome." Feature briefs about a variety of world tourist destinations, accompanied by large photos, were linked by a similar theme: security.

The roundup feature has the perfect structure for online publication: categories of information are already segmented into self-contained stories that link logically to an overall topic. As an experiment, my students and I formed a writing team and agreed on an assignment for a travel roundup: "Top Ten Reasons to Visit Downtown South Bend." Our field trip (and follow-up visits) provided the backbone for our travel roundup, including photographs we snapped and ten linked stories. Then we spent weeks adapting our material for a Web site. Researching and writing the stories didn't take us as long as actually creating the package. The initial feature for the home page and three of the linked features follow; view the site package at http://saintmarys.edu/~cjohnson.

Students also adapted stories they had written earlier in the semester, intended for print publication, to Web sites they created. Figure 9-1 shows the home page for the online magazine "Journey into the Past," created by Evelyn Gonzales and Megan Gamble. The students chose a theme publication to showcase their favorite features of the semester—stories about war veterans. They imported clip art flags that give the site color and motion (the flags wave on the actual Web site). Using clip art also avoids copyright issues. Links to already existing Web sites were chosen to enhance understanding of their stories. Their "experiment" confirms the Internet's unique ability to allow users to customize packages, link information and create themes. These capabilities go beyond any print publication possibilities.

The Writer Online

Established Web sites may have paid staffs funded by advertising revenue, but few positions currently exist for exclusively online staff writers. Many print publications "shovel" stories onto their Web sites. However, a market for freelance online writers does exist. Here's how to become a freelancer for your favorite online magazine.

Some online magazines have no mastheads with editors' names and addresses. If you're fond of such a publication, do a search to find more information. Network. Use the site's "Contact Us" capability to start a dialogue with the editors. Seek opportunities. Veteran online freelancer Nick Roskelly advises novices not to "worry about your writing or credentials. Write something you like. Most likely there's a place to publish it, a space to be filled. You have to be proactive, put yourself in the marketplace.

Journey into the Past

This on-line publicaton honors those individuals who may not be considered textbook heroes, but have demonstrated in their lives the meaning of true heroism. Our mission is to make sure that different individuals receive the recognition they deserve for their accomplishments.

This month features the following articles:

Unknown Hero: James Gamble

* This feature article honors a Vietnam War veteran. For the first time, he is able to speak of his experience.

I Am an American: Ken Takaki and World War II

* Ken Takaki, a Japanese-American during World War II, speaks of his experiences in housing camps during the war and the U.S. sentiment toward the Japanese.

Other links to historical figures:

Color Lines

Holocaust Prayer

Life Hall of Heroes

Figure 9-1   Online Magazine Home Page

Frequent a site, learn its style, look at the site's advertising to determine its demographics. Become a student of the publication."

In addition to freelance opportunities, the Internet also offers writers the possibility of creating their own sites. This involves finding a host, buying a domain name, buying or hiring software, then building the site. Look for someone to fund the site. Roskelly says, "Have something that people want, an overall concept, a business plan (Who will you target? How will you make the site profitable? How often will you publish?)." Once you've launched your site, you will want backdoor tallies of hits to individual stories and to the site itself, information needed to sell on-site advertising such as banner ads.

Writers who create their own sites may develop a community and showcase their own work.

LINKED FEATURE

"No. 1 Reason to Visit South Bend—The Cake Nazi"

Mary Ellen Brown made three trips to Dainty Maid Bakery, located in the heart of downtown South Bend and a local "institution," and no one who worked for the bakery would agree to talk to her. She was ready to abandon the story, but that's not what a journalist should do. When interviews are refused or go flat, a writer needs adaptation skills. Go with the flow; report honestly. When a story is assigned, no journalist should deliberately distort information to make anything or anyone look good or bad. A journalist's primary obligation is to the reader and the truth. After many phone calls, visits to the bakery and research at The South Bend Tribune's library, Mary Ellen told it the way it was. How publishable would this be online? That's an editor's call.

The Cake Nazi. The most affectionate term for the older woman who works behind the display cases filled with freshly baked muffins, cookies and specialty cakes at the Dainty Maid Bakery.

The woman does not wear a nametag, but if you witness her interaction with customers you would soon find out she means serious business. But what if you were not fortunate enough to witness an interaction between the Cake Nazi and a customer and had no idea what you were getting into?

From the outside, Dainty Maid is an inviting, family-owned shop that has been in South Bend for over 70 years. The window displays are lined with red, orange and yellow leaves, adding a festive fall touch. A Barbie doll with a cake dress decorated in white and red icing would be the envy of any young girl. Other fall-decorated cakes fill the window with blank tops waiting for a message.

As I peer through the window on a Thursday afternoon, the bakery looks empty except for three workers standing behind the counter. I have never been inside the bakery but have received a birthday cake from Dainty Maid all three years I have been a student at Saint Mary's College.

You might call it a tradition. My mother and her sister both attended Saint Mary's in the 1970s. They'd ride the bus or walk downtown with their friends to shop, then stop at Dainty Maid to buy a cookie or muffin before going back to campus.

"My first birthday party was in my freshman dorm room in Le Mans Hall," my aunt, Teresa Long, recalls. "My roommates surprised me with a Dainty Maid cake. It was my first cake, too, because I always wanted a pie on my birthday when I lived at home. From that moment, we had Dainty Maid for every occasion. My friends and I even ordered a cake when we were on campus for our reunion a few years ago."

Once you've taken a bite of a Dainty Maid cake, you will instantly wish for another. You won't be able to pinpoint that single ingredient that tugs at your taste buds, but that won't stop you from trying. Is it vanilla? Nutmeg? The cake choices are simple—white, chocolate or banana—but that doesn't make it any easier to figure out what is in the recipe.

But I wouldn't recommend you ask. That would require a confrontation with the Cake Nazi. Yes, the petite woman with salt-and-pepper hair, who wears thick eyeglasses and can barely see over the display cases of sweets. If you do not order, pay and leave the bakery without asking questions or taking up more than a few minutes of her time, you will regret going into the bakery.

And if you happen to be in the bakery when no customers are in sight, and the Cake Nazi says in a matter of fact tone, "As you can see, we are really busy right now," or if it's a half an hour till closing and she says, "We're getting ready to close now" in an attempt to rush you, don't be discouraged. Don't come out until you get what you wanted. And keep eye contact with her. Otherwise she knows you are weakening, and you can't let her think she has the upper hand.

But walking on eggshells is a small price to pay for an unbelievable cake.

LINKED FEATURE

"No. 2 Reason to Visit South Bend—There's No Right Way"

From the backseat of a Toyota, Sara Pendley watched her professor pull up to and get hung up on an old-fashioned high curb in downtown South Bend's shopping district. She observed street and traffic signs as they drove up and down, up and down, the endless succession of one-way streets. She went back later and followed some of the streets to their origin and end points. Then she wrote this story.

Not only is "South" Bend in northern Indiana, but the street names deny the location of the city as well. Michigan Ave., St. Joe St. and Niles Ave. all are named after Indiana's northern neighbor, Michigan. Furthermore, the entire area has been nicknamed "Michiana." Is this an identity crisis?

A parasite of a city, Notre Dame, Ind., feasts on South Bend's identity. This speck of colleges and universities in the midst of South Bend, all of which are dominated by the University of Notre Dame du Lac itself, have overtaken the reputation and tourist industry of South Bend. The sounds of the Irish echo down each street; if only beloved Ireland could hear the call. Once again, the identity crisis comes into play. South Bend has adopted the Irish theme of its neighboring community with Irish pubs and restaurants that name dishes after famous Notre Dame football legends. South Bend is not Notre Dame but strives to grasp the energy, popularity and prosperity that flows from the stadium.

The confusion continues within the city's streets. Visitors must crawl from corner to corner, taking note of teeny tiny "One-Way" arrow signs. Successful at playing "Where's

Waldo?" If so, you'll be an expert at finding the arrows hidden below the street sign names that hang high above the streets, next to the traffic lights. You will never know which way is the right way since one-way streets are sporadically interspersed with two-way streets. You may travel down a two-way street that at the next intersection becomes one way; of course, it is not the way you are traveling. There's only "one" way in South Bend; it's our way or the highway.

Have a compass? Bring it along to navigate down South St., which runs east to west. Also, fluidity is not always consistent. Jefferson Blvd. is an example of South Bend's case of multiple personalities. Through its stretch across downtown, the boulevard becomes Wayne St., going east, and Western Ave., going west. Boulevard to street to finally avenue, this single stretch of pavement leaves you at a loss for names. A traveler must jump a block north to continue down Jefferson Blvd. If you run into a dead end and still have not found the building you're looking for, don't worry. Streets tend to stop abruptly, but they usually restart somewhere.

Parking downtown can be a challenge. If you find a space on the Michigan Ave. strip, a word to the wise. The streets and signs may not reach our standards, but the curbs stand high. Drivers beware: watch for bruised bumpers. If a space is not available, seek a parking garage. The central Michigan Ave. garage is open to the public.

While traveling through South Bend, forget the road map. Bring the compass, a sense of humor and patience.

LINKED FEATURE

"No. 10 Reason to Visit South Bend—You Can Fiddle Away at Fiddler's"

Renée Donovan
*Author of "You Can Fiddle Away at
Fiddler's" as well as "Men's Pickup Lines,
Then and Now" (Chapter 2). Photo cour-
tesy of BreAnna Lewis, St. Charles, Ill.*

As she was on her way to cover another downtown South Bend business, a renovated storefront caught Renée Donovan's eye and made her go inside to investigate. In contrast to Mary Ellen Brown's experience, Renée was warmly greeted by the establishment's owner who abandoned his apron and work in the kitchen to sit down and chat with her. Her feature's tone reflects a very different research experience.

On a blustery autumn day, the flags of the seven Celtic nations beckon a closer look inside the Fiddler's Hearth at 127 N. Main St. Imagine a timeless pub in the middle of "Brigadoon." The bright primary colors of the "proper public house" facade contrast the rest of dreary downtown South Bend.

Step inside the time capsule: a blazing fireplace to your left, a fiddle on the mantle. A smiling staff greets guests and takes them to their own cozy nook. Regulars have already contributed to the decor—family photos, coats of arms, pews from St. Patrick's Church—even though Fiddler's opened recently on Oct. 1, 2002. Carol and Terrance Patrick Meehan saw an opportunity to build something new and took a chance on the purchase of the one-time site of Mickey's Pub, Copper Kettle, Gurbi's and Club Letto.

"If you build it, they will come," Meehan proclaimed during an interview. And he is right. The pub is "becoming home." The Meehans have added their own family treasures and mementos to the decor.

Every great party has great food, and the Fiddler's Hearth is no exception. The rich scent of authentic Irish and Scottish food lingers in the air. Company can enjoy an appetizer of Scotch Egg before a Celtic tradition of Welsh Rarebit, Bangers and Mash, or a Scottish Bridie, among other treats. For the not-so-adventurous guests, regular old American food can be custom made.

Fiddler's Hearth wouldn't be complete without its extensive selection of imported and domestic beers, ciders and spirits. "Proper teas" are served daily from 2 to 5 p.m.

The public house holds a packed crowd of graduate students and alumni of the local universities as well as downtown business people and local families. With advertisements in local newspapers, late night hours and live entertainment, Fiddler's Hearth has become the gathering place.

An Irish pub is a "gathering spot to tell stories, eat, drink and be merry," says Meehan. He describes himself as the "host of the party." The Meehans' son, champion bagpiper Sean Meehan, sets Friday's mood with his 5 p.m. tunes. "You worked a good week so you come to the public house where we say, 'Thanks,' " Meehan said.

Local and international bands entertain into the night. Wednesday night is " Open Mike." Anyone can sign up for a slot to share musical talent, storytelling or poetry, but that doesn't stop guests from getting up on a table and dancing a jig any night of the week.

Chat Room

1. Bring to class a printout of a home page for a magazine or newspaper Web site. In groups, share your Web sites. Discuss the way information is presented. How do you think this layered presentation affects the way you expect to receive information?

2. Do you find shovelware on these sites? If so, how do you feel about it? How could the material be written more appropriately for the Internet?

3. Do you think the Internet has changed U.S. culture? Elaborate on this, then evaluate both positive and negative effects.

4. Log on to America Online, Netzero or another server. What information categories do you find on the initial page? What do these categories tell you about the typical Internet user?

5. Read "The Cake Nazi" again. Does online publishing have looser rules about source identification than print publications?

Create

1. Choose your favorite features of the semester. By yourself or with a partner, create an online magazine to publish your stories. Package the information with as many layers as possible.
2. Form a writing team and decide on a roundup topic (for example, your team might write linkable stories about a recent national, local or campus event or the top reasons to visit your campus; you might review pizza places in your area).
3. Take a shovelware story from a newspaper Web site and rewrite it as a nonlinear story. Package the story and include links to existing, appropriate Web sites.
4. Write a story about a style trend or trendy apparel brand, dating behavior or typical college cliques; use a search engine to locate related Web sites and cite several of them in your story. Refer to your AP Stylebook's "Internet Guidelines" to cite correctly.

INTERNET FEATURE

"Smiling in Terror: An Interview with MC Paul Barman"

Nick Roskelly studied journalism in college, worked for a small newspaper, then joined Chicago's famed Second City writing program. Currently, he divides his time between his work as an associate editor for Stagnito Communications and as assistant music content editor for Static Magazine (http://staticmultimedia.com). He breaks from tradition by writing the following personality profile in first-person voice (see his Click Here in Chapter 11). He believes that style, flair and attitude differentiate an online magazine writer and that it is increasingly important for a print writer to also use first-person voice.

Yazan was the first to meet us. As MC Paul Barman's tour manager and DJ on a short tour through the Midwest and Southeast, "Yaz" is also Barman's only real companion on the road, not counting members of other touring bands like Christiansen, with whom Barman has played several dates with on the tour stretch. Yaz, looking a bit disheveled (he and Barman have just driven from New York City to Chicago and he's coming down with a cold), greets me wearing khaki cargo pants and a sweater. He's got longish black hair, a beard and glasses.

Appearance, schmappearance, right? Typically I have little interest in the way someone dresses, especially in the entertainment world, where threads, scents and shades speak louder than the individual donning them. It's true in all music genres. Even in the nonglitz indie rock scene, a certain protocol for dress exists. And yes, that also includes the hard-to-pindown industrial-trance-funk-jam-yodel scene, which over the past few years has managed to construct a code of its own. However, I noted Yaz's appearance mainly because he seemed to align himself more with my roommates' style than someone affiliated with a fresh and potent hip hop emcee.

A short conversation and then: "Here's Paul."

I turn around to say hello and catch a hooded subject spinning away through a doorway. I look toward the opening thinking: Should I follow?

I'm puzzled. The hooded character surprises me by re-emerging from another doorway behind us. "Hi, I'm Paul," he says pulling his hood down. We try exchanging brief pleasantries, but can't hear each other as Parts and Labor, one of three opening acts, sound checks. We decided to adjourn downstairs to talk further.

An interview typically follows a format. The format adapts to circumstances in a given situation, but it's a format nonetheless. It usually goes like this: interviewer asks question, interviewee answers. With room for tangents, comments and off-the-record rants, the format then becomes cyclical. Following Barman to the basement of the Empty Bottle, I had no reason to think that our interview would go any differently.

But it did. Barman talked openly about several subjects: marriage, summer camp and Tom Hanks' appearance on Family Ties, to name a few. You know, episode 36? The one where Alex learns that his idol and uncle Ned Donnelly, played by Hanks, has a serious drinking problem? So much so that in a frustrated fit of being in a household void of booze, he throws open the kitchen cabinet and pounds a bottle of vanilla extract in front of the Keatons?

He wanted to discuss these topics and a host of others. He was excited about talking. But a question and answer session wasn't what he had in mind. He said things like, "I'd rather just have a conversation," and "Let's just hang out."

And that's what we did. I suppose the traditional format of Q&A seems a bit stilted to Barman. I agree with him, though a charismatic interviewer, ahem, should be capable of putting an interviewee at ease.

The concept of being at ease and MC Paul Barman is a paradoxical mix. On the one hand, he's engaging and genuine in conversation, but on the other, he's uncomfortable with the idea of being interviewed. He relishes his time onstage, but cowers when faced with a camera lens. And he's an Ivy League–educated Jewish kid who has carved out a space for himself in hip-hop culture. "Yet completely alone in the hairiest industry known besides the mafia," Barman utters on "Old Paul," the chorus of which runs: "Old Paul gave rap a cold call/The Caucasoid had the whole block annoyed/It took big gilded gold balls to smile at terror and trial and error." Yet despite all this, his place in the hip-hop community is the one thing about which he seems completely at ease.

Two days after his performance at the Empty Bottle, Barman told me that those times when he had to "smile at terror" mostly centered around his first performances. "I think any time you're doing something new with high stakes [one gets apprehensive]," says Barman. "But I feel like that's all in the past. I'm looking back as Old Paul. Now I can't wait to get onstage."

He has a right to feel that way. The crowd at the Empty Bottle received Barman as an emcee, an entertainer, a comedian and a philosopher. With Yaz spinning records behind him on a bowed buffet table and occasionally playing acoustic guitar, Barman made the crowd dance, listen and laugh. I don't know that I've ever been to a show, much less a hip-hop show, where the audience audibly cracks up. They smiled, placed palms over their mouths, slapped each other on shoulders and pointed at Barman in approval, and became attentive as he rapped "I'm Frickin' Awesome" while sketching a portrait of two girls he brought onstage. They listened to him talk about palindrome research on the "Inter-Web." They even held on when he launched into a lesson (complete with black Sharpie diagrams) about the fourth dimension, which ultimately explained why MF Doom is a slightly better storyteller than Virginia Woolf.

The audience he targets filled the Empty Bottle to capacity, and to follow through with a horrible metaphor, he sank an arrow tip into the bull's eye that is them. And how did he do it? How does he continue to draw in listeners? Wordplay? Check. Humor? Check. Intelligence? Check. Sure, all of these aspects play a part in Barman's appeal, but his essential intoxicating element, I think, is the context he's created for himself. Rather than operate within the guidelines hip-hop has set for itself in terms of subject matter, style and vocabulary, Barman chooses to work through them, or on their periphery, and sometimes in no relation to them at all. From his dislike for interviews and cameras to his choice to go for the laughs in lyrics, Barman seems to maintain a grasp on his impulses rather than assimilate with those his hairy industry sets forth for him. Physically and thematically, he doesn't fit hip-hop protocol and he knows it. But he smiles at terror and moves ahead in his own way.

Chat Room

1. Consult the AP Stylebook's "Internet Guidelines." What is the Associated Press position on the use of lengthy URLs in journalism?
2. What are the differences between search engines and megasearch engines?
3. In a group, generate story topics, then make a list of key words you'd use in an Internet search.
4. What percentage of the Web's information do you think most search engines access? What is the AP Stylebook's estimate? Discuss ways to expand an Internet search.

Create

1. Choose a key word from your list (exercise 3 above) and conduct an Internet search for the word. Use more than one search engine, a megasearch engine and a newsgroup. Share what you find with your group.
2. Create an Internet package to accompany Nick's interview with MC Paul Barman. Include text, photos, graphics, sound and animation.

View

The Internet began in the 1960s as a way for the U.S. Defense Dept. to assure official communication in the event of war. It was not until the early 1990s that the World Wide Web became a social and commercial venue for faster, freer information exchange. As we transition into the 21st century, online news sites are contenders among media industries. Internet stories ideally take a nonlinear structure that defies centuries of storytelling tradition (based on the need for a defined beginning, middle and end). The Internet allows for unique multimedia packaging of stories. Feature writers have become pioneers, exploring ways to respond to the possibilities of the Internet and the needs and expectations of Internet users.

Help

linear structure the typical organization of media stories, moving the reader from a beginning to an end in a sequential or linear fashion.

nonlinear structure a story organized in self-contained "chunks" with embedded links or multimedia elements such as text, photos, graphics, sound, video and animation.

online package various platforms and links utilized to create different levels of information such as headlines, teasers, abstracts, links and interactive activities—chat rooms, polls, games and quizzes.

photo gallery a series of photographs or other graphics posted by an online magazine publisher for site viewers. The quality of Internet reproduction exceeds print, making it attractive for advertisers and publishers alike.

pixel "rows of lighted dots that make up the picture image" in a television or computer screen.[12]

roundup feature self-contained stories written around a central topic, then published together. The roundup is especially suited for online publishing.

shovelware stories from print publications placed on the publishers' Web sites without being reconstructed in nonlinear fashion. Eventually, the practice of "shoveling" stories from one medium to another is expected to become obsolete, with print stories increasingly reconstructed for republication online.

teaser a short verbal message designed to keep viewer/reader interest, common in print, broadcast and online communication.

Notes

1. Sven Birkerts, "Into the Electronic Millennium," *Language Awareness*, 7th ed., ed. Paul Eschholz, Alfred Rosa, and Virginia Clark (New York: St. Martin's Press, 1997)393.
2. Birkerts 393.
3. Alice Z. Cuneo and Tobi Elkin, "E-tailers to See Slower Holiday Growth," *Advertising Age* 12 Nov. 2001:50.
4. Howard Fineman, "The Brave New World of Cybertribes," *Newsweek* 27 Feb. 1995:30–33.
5. Adrienne Mand, "Orbitz Stays on E-marketing Course," *Advertising Age* 29 Oct. 2001:34.
6. Mercedes M. Cardona and Bradley Johnson, "The Ad Market," *Advertising Age* 12 Jan. 2004:8.
7. Tobi Elkin, "AOL Music Deal: Customized Britney," *Advertising Age* 1 April 2002:16.
8. Birkerts 393.
9. Carole Rich, *Creating Online Media* (Boston: McGraw Hill, 1999) 276.
10. Scott Donaton et al., "The Toughest of Times," *Advertising Age* 22 Oct. 2001: S-12.
11. *The Associated Press Stylebook*, ed. Norm Goldstein (Cambridge: Perseus Publishing) 125–132.
12. Rodman 239–240.

10

The Public Relations Feature

Intuition will not suffice.
—*Thomas Bivins*, Handbook for Public Relations Writing

Will I be able to find a job after I graduate? This has become a major concern of today's college student, especially in the tightening job market. Some students' parents worry that the student has not chosen the "right," that is, the practical, major. My own father, a design engineer, lamented that I chose an undergraduate major in English writing. "What can you do with that?" he would ask me.

Indeed.

My university's placement office called me before I had even started to look for a job. The office had received a most interesting call from a local recreational vehicle manufacturer. The company sought someone with a background in creative writing and journalism who could write public relations features for them. At the subsequent interview, I discovered just how unique this position would be. The writer would be given an RV, sent out on trips, and then write magazine features, intended for trade publications, about his or her adventures.

Well, I didn't get the job, but the prospect of paid-for travels in an RV fired my imagination. I eventually did work in the public relations field and discovered that what all the PR textbooks say is true: Entry-level work may involve as much as 90 percent writing.

The formats for public relations writing vary widely. The litany of public relations writing types includes copy for brochures, flyers, newsletters, advertising and annual reports as well as backgrounders, biographies, speeches and presentations, television and radio scripts, and public service announcements. The most common public relations format is the press release. Each day newspapers receive hundreds of news releases from agencies, corporations and not-for-profit organizations. In addition, these same organizations disseminate hundreds of feature releases to newspaper and magazine editors.

The potential for much mobility between journalism and public relations continues to exist. The goal of the public relations writer is to disseminate the employer's message via internal (newsletters, flyers, annual reports) and external (mass media) channels of communication. Who knows the mass media better than someone who has worked on the inside? It is no accident that a number of recent presidential press secretaries were formerly print journalists and television producers. While public relations specialists have tended to earn more than journalists, median annual earnings for public relations specialists fell from $53,312 in 2001 to $47,840 in 2002, according to the Bureau of Labor Statistics.[1] Of course, public relations remains an attractive field for the experienced journalist to enter.

It also benefits the newspaper or magazine editor to know more about the people who send these hundreds of releases. The releases public relations writers send should read exactly like the copy written by the publication's own reporters. This means the public relations writer must write clearly and concisely, using Associated Press style. Clearly, the public relations writer's purpose and process will differ somewhat from the journalist's. Whether you eventually work as a journalist or as a public relations professional, knowing more about each field can only be an enhancement.

The Purpose of Public Relations Writing

Although the practice of public relations takes place largely in the business or corporate world, the public relations writer must thoroughly understand the media and be able to "interview, gather and synthesize large amounts of information, write in a journalistic style," and produce copy on deadline.[2] Writing remains the primary entry-level skill for the profession. A main goal of public relations writing is to secure third-party endorsement through the dissemination of press releases, fact sheets, backgrounders, feature stories, pitch letters and other writings that consolidate key information regarding the organization.[3]

Unlike the advertising message, which the public understands to be biased information from the company that has paid for it, information that appears in an objective newspaper story tends to be viewed as credible. Therefore, the purpose of the writer of the public relations feature release is to disseminate information about her organization, including its history, products and people. The release should be written in the style of the targeted publication. The writer's purpose is not only to get the story published, but also to get it published as is. Most publications adhere to Associated Press style. Public relations writers refer to "The Associated Press Stylebook" to make sure their stories are written correctly. When a story requires much revision, editors may feel they are too busy to rewrite the copy and simply throw the release away. Furthermore, when editors change words, the changes may impact the story's intended message. The fewer edits in a published story, the better for public relations.

Some public relations writing may be published by the organization itself instead of being released for media publication. Such publications include employee newsletters, annual reports and magazines published specifically for employees and members. Sarah K. Magness wrote her feature, "John of the Archives," for the college yearbook. A yearbook records a specific year of campus life primarily through photographic images, so copy tends to be shorter than most newspaper or magazine features. Yearbook layout pairs photographs, frequently clustered in related groups or enlarged to cover full pages or two-page spreads, with brief copy that gives a context for and additional information about the places, persons or events pictured. The yearbook performs the public relations function of communicating positive information about the school to its diverse publics—current students, alumnae, faculty and administration.

Public Relations Proposals and Research

As Thomas Bivins said in the prefatory quote, "Intuition will not suffice." The backbone of the public relations feature is planning and research. Of course, that is the case with all feature writing. However, the public relations writer must plan more formally. The writer typically prepares a story plan, often communicated as a direction sheet, and then submits that plan to her client or employer. If the story seems to be on target and should fulfill the client or employer's purpose in having the story written, then the writer gets the green light to proceed.

"Tune into Reality," a feature release published in this chapter and written by Kaitlin E. Duda, performs a public relations function for MTV. Therefore, Kaitlin added another step to the writing process. Before she even began to research her subject, she wrote the direction sheet (see the Click Here on p. 172–173). Notice that Kaitlin's trend feature appears in feature release format. Newspapers and magazines frequently publish trend features because timely topics tend to interest readers, and the trend feature reflects a norm, product or behavior currently prevalent among a large public constituency.

Types of Public Relations Features

Public relations professionals mediate between an organization and its publics, whether internal (employees, stockholders, members) or external (customers, potential customers, the surrounding community). A large part of that mediation involves written communication, and public relations writing can take on a variety of formats, depending upon the message, the audience and the purpose. Yes, the press release is

Click Here: Public Relations Speechwriting

Public relations writers may be asked to craft speeches for company executives or representatives. Like all feature writing tasks, the process begins with an idea. Who will be addressed? Analyze the audience. What would resonate with this audience? What would not? Do the research. Write eloquently, and don't hesitate to use appropriate quotes garnered through research. Pay attention to tone. President Reagan's famous speech eulogizing those lost in the space shuttle tragedy avoided, wisely, his usual propensity toward one-liners and humor (he had excellent speech writers). In other situations, a speech with a sense of humor might be just the ticket. Former First Lady Barbara Bush became a popular commencement speaker because she knew how to convey a serious message, tempered with good humor.

Here's a bibliography of links to various types of speeches, annotated by Kristina V. Jonusas:

- http://historychannel.com/speeches Here you can easily pick a speech according to a specific topic.
- http://pbs.org/greatspeeches and
  http://pbs.org/greatspeeches/timeline Both sites offer popular, well-known speeches.
- http://gos.sbc.edu This site focuses specifically on speeches given by women.
- http://cs.umb.edu/jfklibrary/speeches.htm Enjoy the Kennedy collection—speeches given by JFK, RFK and Edward Kennedy.
- http://american-comfort.net/speeches.htm and
  http://speeches.commemoratewtc.com/giuliani These sites offer speeches delivered during the aftermath of 9-11.
- http://whitehouse.gov Here you'll find speeches and addresses from the White House.
- http://whitehouse.gov/firstlady This site features speeches and addresses given by First Lady Laura Bush.
- http://dir.yahoo.com/Education/Graduation/Speeches Link to this site for famous commencement speeches. Some of these are really good.

the staple of the PR cupboard. But few public relations writers churn out only press releases. Most, at some time or another, will find it necessary to write in a combination of other formats, a number of them feature formats.

- *Newsletters:* The newsletter became a popular morale booster for organizations, especially manufacturing plants, during World War II. Today newsletters continue to serve the purpose of informing employees about events, benefits and policies in addition to creating a sense of "family." The mainstay newsletter feature is the personality profile, which may be written about employees at all organizational levels. Many newsletters also include shorter articles that focus on everything from minutiae, such as employee bowling scores, to main events, such as births and deaths.

Click Here: Direction Sheet

Advertising and public relations agencies employ people with creative minds and writing talent. Typically, agencies call these people "creatives" and, typically, they work in teams. These teams craft *creative briefs* to share their writing plans with others within the agency or with their clients. The document may also be referred to as a *creative* or *persuasive platform*, although in public relations it is most often referred to as a *direction sheet*. While the name, format and specific content may differ from company to company or agency to agency, the document typically includes an analysis of the audience, the purpose of the writing and the strategic plan for the project. Freelance writers (who query magazine editors with a story) and staff writers (who propose story ideas to their editors) go through a similar process of analysis and prewriting, although the mental work is not necessarily written out as formally as it is in a direction sheet.

Kaitlin E. Duda wrote this direction sheet before she started to work on her trend feature. Getting these ideas down on paper helped her better understand her writing purpose. It also forced her to identify the topic, angle and research possibilities for the story. Kristina V. Jonusas contributed research to this Direction Sheet.

Think of the direction sheet as a directional map for a writing journey.

Direction Sheet

Prepared by: Kaitlin E. Duda
For: MTV

Client Problem:

With a 17 percent increase in its prime time ratings in spring 2000, the launch of its reality shows gave MTV an unprecedented ratings boom. Nielsen Media Research reported that the largest number of viewers in the network's 20-year history, an average of 429,000 people, watched MTV in the fall of 2000. Following the tragedies of Sept. 11, 2001, the network worried that ratings might plummet for its increasingly popular reality shows such as "Real World," "Road Rules" and "The Osbournes." The network feared that viewers might have had all the "reality" they could handle.

Objective:

To position the reality show as a contemporary television staple

Audience Characteristics:

While MTV audience demographics have traditionally been dominated by 12- to 17-year-olds, the reality show has drawn 25- to 34-year-old viewers, an audience that The Wall Street Journal reports "previously shunned the network" (Sally Beatty, "MTV Ratings Soar Off Gross Humor, Sex—And That's Just the Tame Stuff," April 20, 2001). Nielsen reported a 33 percent surge in that age group over the five-year period 1995–2000.

Young people tend to be media savvy and skeptical of marketing. Blatant emotional ploys don't easily work with these generations; they respond better to factual information.

MTV needs to maintain the interest of its viable new viewer mass in the 25- to 34-year-old age range, while increasing interest among its younger audience members 12 to 17 years old.

Creative Strategy:

The communication channel to reach readers will be a feature story about the reality TV trend.

Topic:

The topic of the feature will be the future of reality TV following the Sept. 11 tragedies.

Research Plans:

Research will include interviews with people in the targeted age group as well as a search of Internet sources on the topic.

Style:

I hope to call attention to the reality show and remind viewers of their interest in it through presentation of statistics and facts. The tone of the feature will be objective and professional in order to gain credibility with the targeted readers. The human interest angle will be that we are all curious about what attracts people to certain TV programs. Readers are familiar with reality TV shows but may not know much about how they are created.

Timeline:

Story idea and direction sheet approved by Sept. 15

Complete preliminary research by Sept. 17

Complete interviews by Sept. 18

Complete a draft by Sept. 22

Finalize the story by Sept. 30

Send the story out to the media no later than Oct. 1

- *Annual reports:* Publicly traded organizations are required by law to provide an annual report to all shareholders. To offset the potential boredom of pages and pages of the requisite financial figures, the annual report customarily includes human interest features on new products, product lines and individuals in the organization. Lots of white space and visual support, especially photographs of people featured, make the annual report a more inviting read.
- *Backgrounders:* The backgrounder, quite literally, backgrounds an organization, a person in that organization, a product or an event. Even if the backgrounder profiles a product, it must have a human-interest angle in order to engage the reader through several pages of copy. It begins with a feature lead and may employ direct quotes and anecdotes to keep interest. The backgrounder may be written for internal publics (for instance, to inform shareholders about a new product) or external publics, often accompanying a press release in order to give editors and reporters more in-depth background to a shorter story.
- *Biographies:* The biography or *bio* summarizes facts on key people in an organization. As opposed to the straight bio, which presents "just the facts, ma'am,"

the narrative bio "gives spark and vitality to the biography to make the individual come alive."[4] The feature treatment given the narrative bio results in an informality and flair that makes it appropriate to be used as a speech of introduction in addition to its primary purpose—to provide the media with biographical info on well-known people. The media keeps bios on file for use in the event of breaking news.

- *Opinion pieces:* Many organizations ask their public relations specialists to monitor the media and respond to negative press about the organization or issues of importance to the organization. For example, Barbara Comstock, director of public affairs for the Department of Justice, responded to a Time article, "The War Comes Back Home," in the magazine's June 2, 2003, "Letters" section. She wrote a point-by-point refutation of allegations the Time story made about post–Sept. 11 Justice Department handling of illegal alien cases, representing her organization's opinion. Frequently, publicists for animal rights organizations express their organizations' views in "Letters" sections.

The Release Format

The feature release should be typed on the letterhead of the organization issuing the release. Headlines are written to gain the interest of the media editor and let the editor know what the release is about. The release should be typed and double spaced, with 1-inch margins all the way around. This leaves room for editors to make their editing marks. A contact name and phone number also prominently appear on the first page of the release so that an editor can easily reach the right person to ask questions, verify details or obtain further information.

Subsequent pages of a feature release are typed on plain paper, not on company letterhead. In the top left margin, a slug with a key word from the headline should appear, followed by "add one" for the second page of the release, "add two" for the third page, and so on. Pages should not be stapled; editors may need to separate them during the editing process. The slug and numbering system allows editors to identify where pages belong should they become separated from the story. The word "more" appears at the bottom of each continuing page.

Like all newspaper copy, the feature release concludes with one of the following symbols: three number signs (###), (end), or -30-. These symbols signify that the story has ended.

Kaitlin E. Duda's trend feature correctly models these feature release format details.

"Tune In to Reality"

Trend features focus on a current trend or fad. Assigned a trend feature, a writer should consider recent trends, then choose one to write about. The trend feature challenges a student not only to identify a trend and write about it but also to add credibility and news value with statistics and expert interviews. While she was a college student, Kaitlin E. Duda chose to write about the popularity of reality television. She wondered whether the reality of Sept. 11 would be too much for Americans and lead to a decline in reality programming. (We now know that reality shows have continued to be among the highest-rated programs and are cheaper than traditional programming to produce.[5] However, the popularity of reality shows seems to have peaked with "Survivor" during the 2000 season, with 28.2 million viewers, down to 9.9 million viewers for summer 2003's top-rated "For Love or Money."[6])

Kaitlin E. Duda
Author of "Tune In to Reality."

Part of Kaitlin's assignment was to structure the feature as though a public relations office were submitting it to a print publication. Kaitlin models the correct format for a public relations feature release.

MTV Networks

1515 Broadway Fl 31L
New York, NY 10036-5797
(212) 258-6000

Contact: Kaitlin E. Duda
customerservice@mtv.com

Tune In to Reality

NEW YORK, Oct. 30, 2001—TV enthusiasts need only look at the titles for this week's upcoming reality TV shows to see the trend continues. Channels are full of opportunities to tune into reality, with shows such as "The Real World," "The Amazing Race," "Temptation Island 2," and certainly, "Survivor 3." The spotlight is currently focused on reality TV, and it shows no sign of fading.

What makes these programs so magnetic? Why do people flock to the television to watch "real life" when they live it each day? Angela Fox, a senior psychology major at Saint Mary's College and one of the many followers, provides her insight.

"It is a phenomenon that has caught on because people are bored with their own lives and seek entertainment." She continues, "Or, seeing their own behaviors portrayed in the mass media validates their own lives." This so-called "truth," however, may be stranger than fiction.

Take the casting process of MTV's "The Real World," now showing its tenth season. The station first puts out a call to all viewers on MTV.com, saying, "You sit. You watch. You

-more-

REALITY TV
add one

dig the drama. You long for the adventure. You gripe about how perfect you'd be for the show. So audition already!" The selection process is further highlighted on the Web site: Send in a videotape. Come to an open casting session. However, MTV warns that it does have certain goals in mind.

According to the Bunim/Murray Productions Web site, "MTV, like every other television station, has demographics it hopes to target. Our key market is 12–29 with the majority of those viewers being between the ages of 16 and 24. We hope to reflect those viewers when we cast things." This year, MTV expects approximately 45,000 applicants from all casting venues.

For whom are they looking? The Web site explains, "There are many factors that go into each casting decision, and we always look for the most dynamic, outgoing personalities regardless of their race or appearance." Somehow, season after season, MTV pairs extremely unique individuals together; prepare yourselves for the results!

The reality shows take on different formats, but they revolve around two ideas. The first is that of human interaction, as typified by "The Real World." Viewers watch as the participants live and react to situations together. The second is that of the challenge. Shows such as "Big Brother" and "Survivor" position people against one another. Who will be the best? Who will win the ultimate prize? While there are mixtures of the two, as with MTV's "Road Rules," these are the fundamental formats.

For reality fans, none of this comes as news. Situated in front of the television at the appointed times, fans consume themselves with the emotions of that show. Questions persist: *"Will Will win 'Big Brother 2' and the grand prize of $500,000?"*

The Sept. 11 tragedy revealed the incredible strength of this trend.

Many critics and TV executives questioned what effect the attacks would have on television content. The Sept. 24 edition of Advertising Age delved into the question by looking at the cost of 30-second spots on the prime-time schedule. The magazine reported that its annual prime-time network pricing survey had a new leader, "CBS's 'Survivor: Africa,' which is expected to average $445,000 for a 30-second spot, according to media buyers. TV faces a new reality: For the first time, the most expensive regularly scheduled series in the launch of a season will be a reality program." Rest assured, reality TV is here to stay.

For how long, viewers may ask? Who knows! Old favorites like "The Real World" or "Love Cruise" offer all the drama, competition and high emotion (even on the high seas) of reality itself.

###

🖥️ *Chat Room*

1. Kaitlin E. Duda's feature release deals with the world of business, including conglomerates. How does she achieve human interest while not ignoring the facts and stats?
2. Compare Kaitlin's feature release with full-length features you read earlier in this book, such as the James Earl Jones and Tom Wopat profiles. Discuss similarities and differences.
3. The trend feature requires an element of timeliness. How does Kaitlin tie into news events?
4. Read Kaitlin's direction sheet. What are her plans for the story's tone and human-interest angle? Does she succeed?

Create

1. Turn back to the full-length feature stories about James Earl Jones and Tom Wopat. Rewrite them as feature releases or as one-paragraph bios.
2. Brainstorm a list of current trends. Write a feature story about one.
3. Brainstorm a list of past trends that are regaining popularity. Write a story about one.
4. Cast your trend story into feature release format. You will need to identify a company or organization that would be interested in sending it out. Remember that the purpose of such a feature release would be to create public awareness of the importance of a product or service.

FEATURE STORY

"John of the Archives"

After she was assigned the feature article, Sarah K. Magness had to decide what kind of a feature she wanted to do. This decision was one of the hardest parts of the assignment. However, after she narrowed down her options and chose the topic, she had a great time. Of course, she wanted to write about something that interested her. By doing this, the writing became more personal and flowed more easily. To be certain she could focus on all aspects of the interview, such as the surroundings, and really listen and observe, Sarah created a pre-interview outline of questions. The interview was fairly lengthy. Because neither party had a pressing engagement, they were able to sit down and take plenty of time. Sarah recalls that John Kovach made her feel welcome from the time the interview was set up until she left the Archive Center. He really wanted to do the interview, so he was willing to share as much information as she wanted to hear. Sarah believes that people who feel comfortable and want to talk help create better stories. Kovach was not just someone passing on information; he became a character entwined within the story.

The italicized text was added after the second draft.

As soon as the door *to B26 Cushwa-Leighton Library* opens, curator John Kovach appears. The musty smell of old books and paper engulfs the senses, but Kovach, wearing gray slacks and a sweater, extends his hand in welcome, happy to give a tour of the Archive Center and to introduce its inhabitants.

Few people know of this little "museum" located in the library basement.

Many of the Center's statues and paintings were resurrected from storage rooms on campus. For example, a 14th century painting, found in Moreau Hall and moved to the archives, was a gift from a Vatican bishop. Its monetary value is in the thousands, but that is not why Kovach brought it to the Archive Center. His mission is to give antiquities like this a home so their stories can be told to future generations.

An icon of the Virgin Mary rests on a coffee table in the corner. This icon had sat in the registrar's office for years. An icon of the Virgin is a familiar sight on campus, but this statue of Mary is special in that it was created from an ivory tusk. Many boxes on shelves, tables and the floor clutter the back room. These boxes contain the history of the college, including information about former employees and students.

Sister M. Madeleva Wolff, C.S.C., *served as college president from 1934 to 1961.* Her boxes consume the entire top shelf. Seventeen boxes contain information on Dorothy Feigl, vice president and dean of faculty *from 1985 to 1999.*

Kovach is buried in paper. "Even if all the paperwork stopped coming in, I wouldn't be done organizing for two to three years," he says. Even though he has been curator for a short time, he seems to have spent a lifetime within its stacks. His movements suggest that he knows every shelf and every sheet of paper that has passed through his hands over the past four years. He travels throughout the room as though his course is mapped.

In between journals and yearbooks he finds a trophy. In its cup rests a picture of its first owner. Once the pride of the winner of the 1909 tennis single's championship, the trophy has rested in the Archive Center since its discovery in the Angela Athletic Center a year ago.

For Kovach, the work of the archivist goes beyond housing artifacts. His latest project involves more than an inanimate statue. He is trying to solve the mystery of the death of a student named Zellie Selsby. In September 1870, her senior year, Selsby died. There is no record of what went wrong or even an obituary. All Kovach has is her account record. It reports payments for medicine, first a $5 entry, then just a note saying "prescription." The next entry is a notation for flowers bought to place upon her coffin. Kovach also knows that she has a sizable headstone in the cemetery where, otherwise, only Sisters of the Holy Cross are buried. How was she allowed to be buried there and how did she die? These are just some of the questions Kovach attempts to answer during his long hours in the Archive Center.

Some may consider the Archive Center a boring place, full of musty books and boxes. Kovach urges students to step a little closer. They might see the writings of C. S. Lewis, letters written to Sister Madeleva and the journals of Marian McCandless, *a Saint Mary's alumna who founded the alumnae magazine, Courier, in 1966.* These are the instruments that bring history to life.

The college has visible landmarks that represent what most know of campus history. Le Mans Hall with its Gothic bell tower commemorates the four Sisters of the Holy Cross sent from Le Mans, France,

Chat Room

1. Shorter features are especially suited to suspended-interest structure: A lead captures the reader's attention; the story builds on that interest throughout the body paragraphs and into the conclusion. How does Sarah's arrangement of her material fit this formula? Which paragraphs reveal the most interesting examples, anecdotes and details?

2. Longer features may begin with a lead, then present information in peaks (highly interesting examples, anecdotes and illustrations) and valleys (less dramatic material). The conclusion may pick up an idea or image introduced in the lead. Does Sarah's story do this?

3. Sarah wanted to create a nostalgic tone. Did she? If so, discuss how she did this. If not, how would you identify the story's tone?

4. The italicized text shows detail added after the second draft. How important was it to include these details? What do they add?

Create

1. Draw a graph of this story's structure. It should resemble an upside-down inverted pyramid, in other words, a pyramid–the opposite of news story structure.
2. Write a direction sheet for a story about a campus archive or museum to be published in the campus yearbook or newspaper.
3. After you have written the direction sheet, do preliminary research and set up an interview with the curator of your campus archive or a museum. Ask to tour the facility.
4. Write the feature. Structure it as a suspended-interest story.

in 1843 to open the college. Lake Marian has its legendary "kissing bridge." Even the library in which the archive is housed was designed in a contemporary treatment of the prevailing Gothic design of the older buildings on campus. However, the buildings on campus are not the only representations of its rich history. The artifacts and stories preserved in the Archive Center provide a representation just as rich and much more personal.

View

The feature release is the most common public relations feature format. Other public relations feature formats include newsletter stories, annual report features, backgrounders and biographies. The job of the public relations professional is to mediate between an organization and its publics. Public relations features are written by public relations practitioners as they communicate with their internal and external publics. Like other writers, the public relations writer seeks to relate information creatively and thoroughly. The writer plans stories, researches the material and writes in the style of the targeted publication.

Although public relations writers disseminate their stories to the print media, their situation differs from newspaper and magazine writers who freelance or receive assignments from their editors. The public relations writer must consider an additional audience: the client or company for whom the writer works. Direction sheets allow the writer to record the story plan and to communicate that plan to their clients and superiors before the writing process begins. Typically, the direction sheet includes the client problem, how the story will address that problem, audience characteristics, a creative strategy, topic, research plans, intended style and a timeline. These categories detail the reason for writing the story and how the story will be written.

📖 *Help* _____

annual report   a publication whose primary purpose is to release statistics about the organization. The public relations writer contributes feature stories to add human interest; feature topics typically include new products, product lines and individuals in the organization.

backgrounder   a full-length story that backgrounds an organization, a person in it, a product or a special event.

biography   a feature, commonly known as a *bio*, that summarizes key facts on key people in an organization and may employ details to create human interest.

feature release   a feature story disseminated to the press for free use.

newsletter   a regularly published journal, modeled on the magazine format, used by organizations to inform employees and members about events, benefits and policies as well as to create a sense of "family."

news release   a press release written in news style using the inverted pyramid structure.

opinion piece   a guest viewpoint or letter to the editor of print publications written by public relations professionals to respond to media coverage and/or to convey the opinions or points of view of their organizations on issues of current controversy in the press.

press release   copy written by public relations professionals and disseminated for free use to the press; thus, a news or feature story released to the press may be designated as a *press release*.

public relations   a field in which practitioners mediate between an organization and its publics.

public relations writer   a professional writer with a thorough understanding of the media and a reporter's ability to interview, gather and synthesize large amounts of information and write in a journalistic style. The writer seeks to secure third-party endorsement through dissemination of writing in a variety of formats.

public relations writing   press releases, fact sheets, backgrounders, feature stories, pitch letters and other writings that consolidate key information about an organization. Public relations writing remains the primary activity for entry-level public relations professionals.

Notes _____

1. Daniel Kadlec, "Where Did My Raise Go?" *Time* 26 May 2003:46.
2. Dennis L. Wilcox, Philip H. Ault and Warren K. Agee, *Public Relations Strategies and Tactics*, 5th ed. (New York: Longman, 1998) 12.
3. Monle Lee and Carla Johnson. *Principles of Advertising: A Global Perspective* (New York: Haworth, 2000) 281.
4. Fraser P. Seitel, *The Practice of Public Relations*, 6th ed. (Englewood Cliffs: Prentice-Hall, 1995) 196.
5. Bill Carter, "TV Networks Plan Flood of Reality Programming for Summer," *The New York Times*, 24 Feb. 2003 <http://www.nytimes.com/business/24REAL.html>.
6. Gary Levin, "TV Is on the Brink of Breakout Backruptcy," *USA Today* 9 July 2003:1D.

11

Special Feature Formats

Panning can be fun. . . . But it's also show-offy and cheap. . . . If you really like something, writing becomes humble and stirring.
—Pauline Kael, "Trash, Art and the Movies"

Thumbs up, thumbs down. This was the gladiatorial-style rating system of famed movie reviewers Gene Siskel and Roger Ebert. The two became to movie reviewing "what Arnold Palmer and Julia Child were to golf and cooking respectively."[1] Until Gene Siskel's death in 1999, they clashed in their opinions of various movies on their nationally syndicated television show, making the film review "a spectator sport." While they gained fame as television reviewers, their careers began in print—as critics for two "rival Chicago tabloids."

Siskel and Ebert popularized the movie review, but Pauline Kael (1919–2001) made it an art. For Kael, "films were a passion," and she subsidized her early freelance reviewing by managing a Berkeley cinema.[2] She became a regular film reviewer for McCall's (1966) and the New Republic (1967) before joining The New Yorker staff in 1968. Already 50 years old, she continued with The New Yorker until her 1991 retirement. In the interim she helped launch another famous film reviewer, Paul Schrader, and made her own name a household word as she championed a generation of directors (Scorsese, Altman, De Palma, Coppola and Spielberg).

Kael, like most other successful reviewers, had strong opinions, but these opinions were not arbitrary—they were based on a solid grounding in the area she reviewed. At Berkeley she had studied philosophy, literature and the arts, a background from which she drew her expertise to pass judgment on film artists. And so it is for other great writers of special formats, the opinion pieces and reviews.

For example, Ann Landers pioneered the advice column; Dave Barry and Erma Bombeck brought a sense of humor to everyday life; widely syndicated political

181

columnist James J. Kilpatrick remains chief of the grammar police. Columnists' credentials have ranged from Bombeck's experience as a housewife and mother to the degree in economics that prepared Robert Rosenblatt, recipient of the Loeb National Award for financial journalism, to write his "Health Dollars and Sense" column for the Los Angeles Times.

Writers offer their opinions (based on personal experience and/or formal training) in special formats. While students may find a venue for their personal opinions in a campus newspaper, few novice writers will be assigned to write columns in entry-level jobs. Thumbs up or thumbs down, the opinion piece and the review are the domain of the experienced feature writer, although there will always be some exceptions.

Opinion Pieces

While you might think an opinion piece would be the most enjoyable type of feature to write, my experience tells me it can be the hardest and most painful. Pauline Kael was exactly right in her observation that even though panning something or someone may be fun at the time, it has its repercussions when the pan hits print.

It's also true that newspaper interns and novice newspaper and magazine writers will find opportunities to write opinion pieces few and far between. As is clear from the name of the format, *editorials* are the domain of the editor—people who have a good deal of writing experience already behind them and generally represent the views of the publisher or the views of the publication's editorial board.

However, anyone may take a crack at writing an opinion piece. Newspaper editors customarily share their editorial pages with readers, both *letters to the editor* and reader op eds (full-length opinion pieces). Magazines do the same; for example, Newsweek features a reader-written column, "My Turn." Although many readers send in letters and *op eds*, space limitations dictate that a small percentage will be published. Knowing how to write a compelling opinion piece should increase the chance that the writer will be published.

Opinion pieces should be short, punchy and original commentary on an incident or issue that is both timely and controversial. Like any other feature, the best ones will be backboned with research. President William Jefferson Clinton generated some of the most polarized editorials I've ever read. Like President Ronald Reagan, Clinton was more than an American president—he became an American icon. New Yorker writer Adam Gopnik, in his collection of features "Paris to the Moon," calls Clinton "Barney for adults": "Bill Clinton just held out his arms and watched people leap into them."[3]

When the Monica Lewinsky story broke, it looked as though the United States had turned into a Flannery O'Connor novel. "A Leader Much Like Ourselves" reflects my opinion on this phenomenon and my speculation on the "why" of U.S. reaction. The piece went out on the Scripps Howard news wire and was published all over the country, including in The Atlanta Constitution. I am very aware of the Atlanta publication because the most stunning hate mail I received came with an Atlanta postmark. Note that the piece blends personal with scholarly opinion.

Another op ed I wrote, about the "Seinfeld" TV series, appeared in Vanity Fair. Always include a phone number when you submit writing of any kind to a publication; a Vanity Fair editor called me to ask permission to cut the piece. It was finally so short that it appeared in the letters column. The letter is reprinted in this chapter as it appeared in the magazine. It also generated hate mail, an anonymous diatribe postmarked in New Jersey.

These were the last opinion pieces I've written. Getting your opinion in print may generate a feeling of power, but it also puts you out there with a view someone won't like. Whether you publish your opinions in a newspaper or magazine, prepare for a critical response.

In addition to editorials, letters to the editor and op eds, newspapers and magazines typically publish *columns*, a feature article that makes a regular appearance in the publication. Some columns may be written exclusively for a specific publication, such as Robert Rosenblatt's "Health, Dollars and Sense" column for the Los Angeles Times. Others are written by staff writers for specific publications and then are syndicated, such as George F. Will's columns. Will is a regular columnist for the Washington Post, but these columns appear in newspapers all over the country as well. Gwynne Dyer, who writes about issues of international interest, is an independent columnist based in London and published in 45 countries. James J. Kilpatrick's columns are distributed by the United Press Syndicate.

Columns might be categorized by rhetorical strategy (humor, curmudgeon) or by theme (advice, sports, world, politics, law, science, technology, social issues); however, many types overlap because these "are skilled writers whose voices find expression in many of the strategies of rhetoric."[4]

Reviews

Reviews are a subset of the opinion piece; they convey someone's opinion about a book, movie, play production, dance or music concert, recording artist or art exhibit. Magazines also publish reviews, but newspapers have traditionally published them more quickly and thus have had the greater historical impact. For example, Frank Rich of The New York Times probably closed more plays than any Broadway producer. Today Internet and televised reviews undercut newspaper's quick response; the famous pair of Siskel and Ebert broke new ground as they left print journalism and broadcast their thumbs up and thumbs down to a nation. However, if you think a newspaper review lacks teeth, try publishing a controversial one and then watch the Letters to the Editor column.

Length differences between magazine and newspaper arts reviews are hard to assess. The dance reviews I published in the regional magazine Arts Indiana ran about 1,000 words, while the arts reviews in the regular "Goings On About Town" section in The New Yorker average 100 words. Newspaper reviews fall somewhere in between.

Whether 100 words or a thousand, arts reviews follow a distinct format. Within the first few paragraphs, reviewers must make their opinion of the work clear to the reader. Reader resource information for the person who may want to view or buy the work should be detailed in the review or a sidebar: the title of the work, when and where a production or exhibit is available to the public, a release date for a recording or film, ticket prices, and Web site addresses and phone numbers.

The reviewer should have some expertise to evaluate the work; that is, he or she should know more about the art than most readers. For example, a film reviewer might watch classic movies, read books about the cinema and learn the vocabulary of movie production. The best arts reviewers generally are those who have had hands-on experience. Someone who has worked on a production knows the amount of time required to mount a play and the importance of the ensemble, not only actors but also other key contributors whose work determines the tone and interpretation of the play—the director as well as the costume, lighting and set designers.

The reviewer needs to convey an overall impression of the work through description and evaluation of its elements. As with all feature writing, the writing should not tell everything that happened or describe everything observed—judiciously selected details should support opinions. For example, I reviewed a dance concert in which the artistic director introduced each dance number over a public address system while the audience stared at an empty stage. "The bare stage," I wrote, "made narrations seem unduly long. Perhaps a dancer or two in silhouette or motion would enhance the narration, which at times should be abbreviated."

Novice reviewers should use adjectives sparingly and tap the power of action verbs. I made the verbs work in an Arts Indiana dance review: "Golden light flooded the stage like a glowing, burning sun." Bells "accentuate" footwork, and movement "bespeaks" human frustration. Mention all artistic aspects that contribute to the whole: "Dancers jitterbugged to Jerry Lee Lewis' explosive 'Great Balls of Fire.' " And strut your stuff, teach as well as inform: " 'Sleeping Beauty' premiered January 1890 at the Marlinsky Theater, St. Petersburg, Russia. The ballet featured Tchaikovky's music and Marius Petipa's choreography. In this production, Petipa's lilting choreography, with its colorful, romantic texture, unfolds like a flower."

FEATURE STORY

"8:16 a.m."

On Sept. 11, 2001, Megan Colvin had already endured a 24-hour fast and planned to write a participatory feature about it. However, she realized a unique opportunity as a writer to "participate" in something that changed the world. She wanted to record the emotion of that day, and the opinion piece seemed a perfect format. The story broke some major rules of feature writing: It has no quotes, comes from a limited perspective and lacks background information. Only when Megan stopped limiting herself to the assigned format did the story start to take shape. Megan offers this advice: While rules are made for a reason, sometimes they simply need to be broken. Her story was published in the December 2001 issue of the Saint Mary's College Courier.

All I wanted to do was check the weather.

When I turned on the television, though, I got much more information than I had ever anticipated. The violent image that greeted me indelibly marked itself in my memory. It was 8:16 a.m., 53 degrees, and two gaping holes had just been torn through the World Trade Center.

The voice of an obviously distressed Bryant Gumbel offered commentary on the fiery scene. Still half asleep and semi-oblivious, I thought the crash had been an accident. Gumbel's voice behind the tape of a plane careening into the second tower snapped me back into reality. He simply stated, "This obviously was an attack of a deliberate nature." That single sentence changed everything. When I heard a plane roar overhead a few minutes later, the familiar sound gave me chills.

I had awakened in a still dark and quiet hallway. Now fluorescent lights burned bright. Frenzied conversations about the disaster were punctuated with ringing phones and the beeping of computers as instant messages arrived. The authoritative drone of assorted newscasters blended together as they told of the unfolding situation. Collective gasps could be heard as the horrific events continued to unfold.

I watched live as the second tower collapsed. In just a few seconds, the skyline of New York City changed before my eyes. This great symbol of American economy simply fell down in a great cloud of dust and smoke. I looked at the clock and saw that I had been awake for just over an hour. It didn't seem like enough time for the world to have changed.

Just after noon, I walked to the prayer service. What had begun as a small group of friends grew to a huge mass of people as we approached the chapel. Entering the crowded room, I saw every seat already filled. I squeezed myself into a small open space on the cold marble of the altar steps and watched people continue to file in.

I had trouble concentrating during the service. I did not pay attention to what was said or what songs were sung. I did not notice how the chapel had been decorated, or even if it had been. Packed so tightly in the chapel, I could only think of one thing. I did not want to be there. I did not want anyone to be there. I was angry that we had to be there. Being there meant admitting the attack had happened and the world had changed because of it. The finality of the situation was overwhelming.

The rest of my day was spent completely immersed in the aftermath of the terrorist attack. I watched CNN on a big-screen television as I ate lunch. News pre-empted music on all the radio stations. Probably for the first and only time, Dan Rather broadcast the news on MTV. It was as though the normal world had stopped.

But it hadn't.

My laundry still needed to be washed, I still had a quiz to study for, and my sunglasses were still nowhere to be found. The fact these thoughts even went through my head made me feel guilty. It seemed disrespectful to think about such trivial things in light of what had happened. But I knew that the next day my world would be back to normal.

There was a changing of the guard of reporters and anchorpersons at what would have been midnight in New York. Some of the faces I had watched since the morning finally said good night. Realizing how long I had watched, I decided to follow their lead and go to bed myself. As I tried to fall asleep, I thought of the horrific images I had seen that day: raging fires, masses of people running for their lives, desperate individuals leaping from the highest floors of skyscrapers. They all seemed like scenes from a bad movie, one of those action-filled blockbusters with incredible special effects and implausible plots. News footage of people running as the towers collapsed was reminiscent of "Independence

Day." The explosions looked like special effects from "Armageddon." But Sept. 11 wasn't the latest big-budget Hollywood production. It was reality.

My alarm went off at 8:16 on Wednesday morning, just as it had on Tuesday. But the days were nothing alike. They stood on either side of the before and after that now divided American history.

I wish no one knew about Columbine High School. I wish no one recognized the name Timothy McVeigh. I wish no one would remember exactly where they were and what they were doing on the morning of Sept. 11, 2001, for the rest of their lives. And I wish my generation's history didn't have the common denominator of violence.

But you can't always have what you wish for.

 ## Chat Room

1. Compare this personal story to "Making Stone Soup" (Chapter 1). Do inner or external conflicts dominate these stories? Discuss the conflicts mentioned in the Sept. 11 feature.
2. This feature concludes with an insight about the current college-aged generation. What is this insight? Discuss your own world views.
3. Megan felt this feature broke the rules of feature writing because it lacks quotes, comes from a limited perspective and lacks background information. What is it that makes this such a successful feature nevertheless?
4. If Megan had written this piece as a column, how would you characterize the rhetorical strategy? The theme?

Create

1. Recall your feelings on Sept. 11, 2001. Interview friends regarding theirs. How do they compare/contrast with Megan Colvin's? Write your own opinion piece.
2. Search the news for other world events that would be conducive to features similar to Megan's. Make a list of possible topics, then do an online search to see whether other writers have chosen your topic and written features on it.
3. Think of an area in which you have experience and/or expertise. Identify a theme and rhetorical strategy for a column.
4. Write a few sample features for your column. Send a proposal detailing your idea to the editor of your campus newspaper (include the samples).

✏️ *Click Here: More on Those Pesky Pronouns*

As both an associate editor for Stagnito's New Products, a Chicago trade magazine, and assistant music content editor for the Web site "Static" (http://www.staticmultimedia.com), Nick Roskelly knows all about writing to the audience. For that reason, he takes issue with writing teachers who routinely prohibit a casual voice, including colloquialisms and the first person. Having to write for extremely different audiences in his current position and during a past stint as a writer for Chicago's Second City comedy revue, Roskelly believes the chance to exercise flexibility in style and voice serves the novice writer well.

Features and reviews can be very different and shouldn't be subject to the same absolute guidelines. To say that neither is publishable if written in the first person creates a sweeping generalization that intelligence warns against. The truth is that a good writer has many voices and stylistic weapons in his arsenal. The best writers know when to use them based on their audience. In newspaper, an entire community, spanning marketing demographics, (hopefully) reads. Writers don't know precisely to whom they are speaking. Both literally and figuratively, newspaper reporters and their audiences are strangers. Outside of opinion pieces, the informal first person has little place here.

However, a figurative friendship exists between magazine writers and their readers. Magazines, for the most part, target specific demographics. A reader picks up Mother Jones because she has a specific interest. Another reader buys Photo because he has another. Writers for these magazines address their readers in a particular way, using appropriate voice and style and sometimes the first person. Occasionally magazine readers want to hear a writer come from his personal perspective during an interview with a subject. Setting the scene of an interview, pointing out small details: shifting eyes, a tapping foot, a grey tooth, a scarred thumb, relaying what a subject asks an interviewer . . . these kinds of elements lend themselves to the first person. The first person gives life to the situation and circumstances shared by interviewer and interviewee; this is a situation that no one else, no reader and, more importantly, no other magazine in the publishing world can re-create. The first person assists in differentiating magazines from one another, and these days that's the name of the game.

Features and reviews share research, critical thinking and good, old-fashioned writing mechanics as essential elements. But adhering to a policy against the first person isn't only counterproductive; it's unrealistic. Too many media outlets exist. Too many voices are out there. Too many opinions worm their way into the world disguised as fact because they weren't written in the first person.

Pursuing an objective goal in journalism should always be on a writer's mind. But in the context of features and especially reviews, a writer may opt for honest opinion rather than muddied objectivism. Allowing the first person gives students a chance to better understand the tone it sets. When they publish, they'll be able to make a more educated decision about whether or not to use it. To tell them to avoid it altogether would do them a disservice because there's a ton of opportunity for writers who bring first person to feature and review pieces.

Opinion Piece

"A Leader Much Like Ourselves"

This opinion piece, released by the Scripps Howard News Service on March 17, 1998, was subsequently published in newspapers around the country. The Atlanta Constitution featured the story in its Viewpoints section and subtitled it, "He who is without sin: The American psyche often can accept moral failings before it can respect self-righteousness."

During an interview for an entertainment story in the summer of 1996, actor Tom Wopat, a TV icon of "Dukes of Hazzard" fame, predicted President Clinton's re-election. "He's the sinner-preacher," Wopat told me. "And America loves the sinner-preacher." If not a preacher in substance, Bill Clinton certainly is that in style.

Meanwhile, thunderclouds of sin hang over Washington.

Paula Jones' fund to pay attorney fees is said to be going directly to her even as news stories have turned to questions about the ethics of the independent counsel, who, in turn, is investigating the White House.

Clinton has withstood years of allegations of immoral and illegal activities of various kinds from his detractors, yet his approval ratings have shot up as high as 73 percent. Prosecutor Kenneth Starr apparently suspects that Clinton's buoyancy results from the work of the president's public relations spin machine. Indeed, he has been hauling the president's men into his inquisition to discover the secret weapon.

Starr might benefit, instead, by looking into an American psyche that he has largely ignored.

In 1996, Wopat said, "Clinton is Elmer Gantry," referring to the sinner-preacher in the Sinclair Lewis novel. "Elmer Gantry" tells of the American heartland's attraction to a wide-smiling, womanizing man who sweeps through the Corn Belt with a thunderous indictment of sin—while doing all that he decries.

Though Lewis' novel was a product of the 1920s, its unfolding of American myth gains new relevancy in the 1990s. Beset by the post-Watergate media, which has bombarded the public with a litany of corruption—corrupted politics, corrupted food supplies, corrupted air—the American public probably has come to the end of its tolerance of resurgent Puritanical oppression. Battered and beaten by intricate knowledge of every evil event, the public may have had enough of the god of wrath that Starr has come to exemplify.

An anonymous electronic message forwarded to me parodies Starr in the language of Dr. Seuss:

"I am Starr. Starr I are.

"I'm a brilliant barri-star."

His self-righteous, pristine tone contrasts the media's image of Clinton as the sort of backwoods Bible thumper satirized in Flannery O'Connor's "Wise Blood." In this work, O'Connor subjects sin to cartoonish exaggeration and paints the weaknesses of the flesh as so pervasive that no one can escape the taint.

The media, too, offer a similar cartoon of official sin, but without O'Connor's theme of redemption. In this re-emerging picture of sin in the landscape of the American mind, no one can escape corruption. It becomes understandable, then, that the public rallies to its sinner-preacher. Postmodern America is no longer disillusioned about the atrocities of its past. And statistics on domestic abuse, divorce, adultery and crime prohibit any misconceptions about our personal lives.

We find catharsis in the fatal flaws of our heroes, but we no longer wish to exile them as the ancient Greeks did. For the ancient Greeks, perfection was possible. For us, it is not.

Few of us could withstand the prolonged scrutiny that Clinton has endured. The plight of innocent, ordinary people in Arkansas whose lives and life savings have become casualties of the Starr investigations have also evoked another sobering thought: Any one of us could be subjected to such invasive treatment.

Scripture comes to mind: "Judge not, that you are not judged. For with the judgment you pronounce, you will be judged" (Matthew 7:1–2).

Clinton's punishment or banishment would alarm the majority of Americans who have come to believe that corruption is unavoidable.

The public's support of its president telegraphs a nation's need for still waters, for the kinder, gentler America that still eludes us.

COLUMN

"Looking to Get Hitched, Just Not Yet"

Feature story ideas can come from any aspect of life. Molly V. Strzelecki says the inspiration for the following story came one night as she and her college friends were "sitting around the computer, bored out of our minds, looking up wedding stuff on the Internet." Molly had just been asked to write a column, "Growing Up to Be a Kid," for the campus newspaper, and this became her first column. As assistant editor of a trade food publication, she still writes columns, now for Snack Food & Wholesale Bakery magazine.

It snuck up on us like a cat through the night. It was reinforced through print, computers, television and all other mediums put together. Once the words dropped from one person's lips, they ran like wildfire through the eyes and ears of the local population, consuming us in a wrath of ribbon, satin, lace and taffeta. We were getting married.

Not any time soon, mind you, but when the four of us first logged on to the Modern Bride Web site, the gleam in all of our eyes could not for the life of us be dulled. We all wanted to get married right then . . . well, not actually married per se, but more we all wanted to plan a wedding of our own, to have all the attention focused on us in our long ball gowns and diamond rings, hair done in pin curls and honeymoon in Barbados. It was set. We dreamily set our sights on a last-minute lipstick touchup, fluff the train, open the doors and . . . ahhhh . . . wedding . . . me . . .

Me? Married? There's a thought that . . . well, yes, I have thought about it, but right now I'm 21. I have always thought that being married by 24 or 25 would be ideal. What I didn't realize, however, is that the age of 24 is oh my God only three years away. I don't even have a boyfriend. I don't even have a prospect. Now, realistically, the young women and even young men for that matter, in my same predicament, could set themselves on a quest for a partner and make this an ordeal, because really, in my mind it is becoming an ordeal, but do I have the time? Frankly, no. At the moment I have tests to study for and errands to run, neither of which force me to compromise my intelligence or emotional capacity, and both of which are a hell of a lot easier than marriage.

But imagine, as I described before, five or six 20-something girls sitting around in jeans and sweats, faces glued to the computer screen, not furthering their education in any way except to learn the many intricacies of certain matrimonial Internet sites and possibly the psychology behind creating ideal wedding situations in one's head based solely on a whole lot of hope. It is a far cry from the bobby socks and ponytail days of yore when girls really did marry Bobby the football captain a week after high school graduation. Nowadays marriages seem to come much later in life and even in multiple numbers. According to the National Center for Policy Analysis (NCAP), 14 million Americans between the ages of 25 and 34 have never been married. And in 1998, 56 percent of U.S. adults were married. A pretty high percentage, right? Not really, considering that's down 12 percentage points since 1978.

It has to be that an unconscious, undiscovered section in the female brain at some point or another clicks on to weddings, then maybe shuts off after a while and possibly (if ever) clicks back on at a later date. I'm pretty sure that this self-indulging wedding talk proves only to satisfy one's ego. Face it, your wedding is a day devoted to you. All eyes on you. Bridesmaids' colors, accent colors, what dish to serve, the script used on the invitations, flowers, the church—it is all up to you. You, as the bride, the mother of this child called a wedding, get to choose every aspect, with some help, of course, from mothers, friends and the husband-to-be. But in essence, you, as the bride, get to play God. "Power unto me!" you may exclaim in a booming voice as the first picture book of wedding cakes is set in front of you.

According to TheKnot.com, 80 percent of Americans consider marriage part of the "good life." And this piece of the "good life" pie will only cost you about $19,000—a mere pittance compared to the fairy tale weddings of celebrities we are bombarded with on television. So not only can thoughts of weddings consume every atom of our beings, but they can also physically consume every aspect of our wallets. Maybe it is because I am neither married nor engaged, but I cannot fathom spending what for me right now is a year's worth of education on a huge one-day party. This is not to say that I wouldn't do it, I just can't see myself doing it right now. Maybe that's because I only have $43.67 in my checking account.

A wedding represents a lot in America's culture: love, friendship, family and a really big party. And there are worse things to which our thoughts could be devoted. The best part about fantasizing about our future weddings, though, is that the fantasy is always a happy story, despite what the reality might be. Now that my friends and I are officially out of our teen years, the pressure is on to find that man (or woman), zip up that long white dress, pin a veil to our heads and march ourselves down the aisle. But as the old adage goes, you can't force something if it isn't right. Marriage is not for everyone. Personally, I enjoy the thought of being married but not right now.

I have heard of a new Web site that can coordinate an entire wedding online with just a date and the husband-to-be's name. The hunt is on.

Molly V. Strzelecki, a senior English major, is a new columnist to the Viewpoint page. Her column will appear every other Tuesday. The views expressed in this column are those of the author and not necessarily those of The Observer.

"Judeo Jerry"

A letter to the editor of a newspaper or magazine, while not a full-length feature story, at best has feature characteristics: the brief opinion piece reflects on an issue usually raised by an article or another letter published in the targeted publication. The successful letter condenses elements of argument: identifies the issue, makes necessary concessions to the opposing view and presents a case with factual details. The following letter appeared in Vanity Fair, July 1994:22.

As one of the first serious scholars of *Seinfeld*, I appreciated Michael M. Thomas' quest ["Air Seinfeld," May] to explain the show's phenomenal popularity. He is certainly correct in suspecting that the show is a "throwback to an earlier age of comedy." However, he has overlooked the show's true roots—not in screwball comedy à la Cary Grant, but in traditional Jewish humor. Thomas touches on this revelation himself when he compares the show to the films of Woody Allen.

In fact, the show's roots stretch back centuries to the Yiddish folklore of the Eastern European shtetl, where the luckless schlemiel suffered alongside his equally luckless sidekick, the schlimazel. Shticks that frequently cast George as the classic schlemiel, the inept bungler who accidentally spills the soup, and Jerry as the classic schlimazel, in whose lap the spilled soup lands, characterizes *Seinfeld's* humor. In his book, *SeinLanguage*, Jerry himself acknowledges his own Jewish father, "a comic genius selling painted plastic signs," as the immediate source of his own comic inspiration.

DR. CARLA JOHNSON
Notre Dame, Indiana

Chat Room

1. Discuss Nick Roskelly's Click Here in this chapter. How do you feel about writing in first person? Do you feel it's a "cop out," that it's easier to write first person than objective third person? Do you think student writers are inclined to prefer writing in first person? Why?
2. How does "Judeo Jerry" compress the elements of argument? Identify its argumentative components as well as the ways it reflects on a feature published in the magazine's May issue.
3. Go to http://columnists.com and click on Hall of Fame to read about major columnists. Search for some of their columns; discuss their rhetorical strategies and themes.

Create

1. Read Flannery O'Connor's short story "A Good Man Is Hard to Find" or "Barney in Paris," a feature published in Adam Gopnik's "Paris to the Moon." Write an opinion piece refuting, qualifying or agreeing with the idea that Bill Clinton is like Barney or an O'Connor character.
2. Write a letter to the editor of your campus newspaper in response to an article it has published. Submit it to the paper for publication.
3. Write a local history column or an opinion piece about a historic event.

ARTS REVIEW

"Some Like It Hot"

Subheaded "American shows popular during London heat wave," this is a theater review I faxed home in the summer of 1995. While in England for a month, I agreed to review London and Stratford theater. Since I couldn't find a typewriter or computer, I wrote the reviews by hand, not my usual composition method. The review appeared on the front page of The South Bend Tribune's entertainment section on Aug. 13, 1995.

Variety, quality, grand surroundings—London theater is hot and, this summer, quite literally. The heat wave you've been experiencing is international, and the Brits are as unprepared as Americans have been.

There's reason to brave the heat for London's HOT theaters which offer everything from "Buddy: The Buddy Holly Story" to Arthur Miller's "A View from the Bridge" (the great American playwright is easier to find this side of the Atlantic).

One of the hardest shows to get tickets for is "Crazy for You," a new version of Gershwin's 1930 musical "Girl Crazy," at the Prince Edward Theater. As Bobby, Tim Flavin, a native of Houston, heads a mostly English cast (their American accents are almost as good as his).

The Gershwin songs carry their weight in nostalgia, but the play is a puff of fluff. The ins and outs of Bobby and Polly's love affair are wholly predictable. But the show dazzles with costumes, well-timed slapstick routines and a joyous cast, including a charming chorus of smiling hoofers.

Helen Way's lanky, lyrical Polly evokes memories of Ginger Rogers in the Astaire-like finale on London's art deco set.

A touring company will present "Crazy for You" at the Morris Civic Auditorium in South Bend in November, part of the Broadway Theater League's season.

More interesting, if less lighthearted, is "Sunset Boulevard" at London's Adelphi Theater. Though most of the show's press has centered on its stars—Glenn Close, Betty Buckley (who left the London show July 4 to replace Close on Broadway) and Elaine Paige (Buckley's replacement), the show may not be the star vehicle critics have thought.

Alisa Endsley takes over as Norma Desmond on Paige's Thursday nights off, and both Endsley and the production present Andrew Lloyd Webber at his best.

Admittedly, Lloyd Webber may be going through a grotesque phase, but "Sunset Boulevard" has substance. In his daring marriage of film and stage, Lloyd Webber recasts the Gloria Swanson–William Holden 1950 classic into a lament for a lost era of Hollywood glamour.

When Desmond's butler–chauffeur–ex-husband says, "We gave the world new ways to dream," he's talking about more than one has-been director and one aging actress.

The tale unfolds with the swelling emotions and cresting moments of Lloyd Webber's music against a set worthy of Edgar Allan Poe. A "handsome Hollywood heel," a down-and-out screenwriter, is drawn into the web of Desmond's desperate fantasy, but the gloom-and-doom darkness assures us that the destruction of these remnants of glamour cannot be avoided.

While Endsley resembles Paige facially, she is not as petite (Paige is so tiny she makes everything else loom larger) and plays Desmond with more subtlety. Endsley lacks a big name but proves that she, and the show, are solid.

FILM REVIEWS

"Chicago"

Marianne Orfanos and Laura Coristin viewed the movie "Chicago," then wrote New Yorker-style in-brief reviews. While much could be said about this complex film, the students were asked to come up with an angle and to keep the focus narrow. Writing short is difficult, an art in itself that requires the writer to keep a spotlight on the central idea.

Marianne Orfanos
Author of "Chicago" review.

If you're looking to reminisce about the "good old days," then "Chicago" is not for you. This tale of glitz, glamour, greed and corruption doesn't paint a picture of a carefree time. But come anyway and join in the debauchery as Catherine Zeta-Jones, Renee Zellweger and Richard Gere "give 'em the old razzle dazzle" in this revival of the Bob Fosse Broadway musical. The inspired dancing and seductive beat will pull you from your seats with songs such as Queen Latifah's torchy "When You're Good to Mama." You'll be shocked at what she can do with a scarf. On the heels of "Moulin Rouge," director Rob Marshall has created another successful movie musical. —*Marianne Orfanos* (1/14/03 Showplace 16)

This is a musical brought to the big screen with all the charm of theater. Catherine Zeta-Jones and Richard Gere return to their musical roots as Velma Kelly and Billy Flynn, proving that the stage is their home, while Renee Zellweger's musical debut as Roxie Hart is promising. Rapper Queen Latifah gives an impressive performance as the "keeper of the keys," Mama Morton. A story about murderous women and their desperate need for publicity—good or bad—"Chicago" is an entertaining production. Director Rob Marshall keeps the catchy musical numbers onstage even though this is film. In the words of Billy Flynn, "That's Chicago." —*Laura Coristin* (1/14/03 Showplace 16)

MUSIC REVIEW

Common's Electric Circus (MCA)

His work as assistant music content editor and staff writer for Static Magazine (http://staticmultimedia .com) gives Nick Roskelly a respite from his work as an associate editor for a Chicago trade publication. The trade publication industry is completely different from the newspaper business, he says. A trade publication's purpose is to write to people who are trying to make their businesses better, with focus on such topics as industry trends and how to make a dollar stretch further. Initially a newspaper staff writer, Roskelly says writing for Static brings him almost full circle, back to his newspaper skills. He sees online publishing as new and open, and he especially enjoys covering hip-hop music for the online magazine.

It took me a little while to lay my hands on a review copy of Common's Electric Circus, and by the time I did, several people had already leaked their opinions of it to me. I hate when that happens. Now I don't know if I'm giving you their review or mine. For what it's worth, opinions of Electric Circus seem to fall into two categories: cool and weird. I think I'd file it under both. Having said that, let's explore what the shallow descriptions of *cool* and *weird* mean on this album.

Cool: The cooing of Jill Scott on "I Am Music" over growling Dixieland horns; banging bass piano notes halting for a groovy chorus with Stereolab's Laetitia Sadier on "New Wave"; a falsetto from Pharell Williams claiming "I got a right ta feel high," on "I Got a Right Ta."

Weird: The cooing of Jill Scott on "I Am Music" over growling Dixieland horns; banging bass piano notes halting for a groovy chorus with Stereolab's Laetitia Sadier on "New Wave"; a falsetto from Pharell Williams claiming "I got a right ta feel high," on "I Got a Right Ta."

Electric Circus is an ambitious attempt at pushing the boundaries of hip-hop into new musical territories. The ambition and exploration on each track leaves a listener thinking, first, "What the hell is this?" and then, "I don't know, but I kind of like it." Cool and weird come together as one. In both the music and the art worlds, cool and weird are typically linked up, though each has components of the thrilling and adventurous. By contrast, in hip-hop, weird almost always yields to familiarity and comfort. Hip-hop typically searches for what's real and then presents that reality in a palpable way. Keep it real. Tell us straight. Don't bother me with hermetic poetry and metaphor. Don't start talking like you know something I don't. Hip-hop is something we all share, and if you think that you're special somehow and better than it, then get out.

There are, of course, plenty of cool weirdos in hip-hop. Andre Benjamin of Outkast, Kool Keith, Busta Rhymes, the Beastie Boys, to some degree, and I always thought Kwaméé was a little strange (all those polka dots and whatnot). But most of these hip-hop eccentrics have innovated within the music's boundaries, lessening suspicion from a hip-hop community ready to ostracize artists who don't play the game as it's traditionally played.

I suppose what's different about Common is that he's an artist who hasn't always traveled down the weird road. He's been accused of being too deep and melodramatic in his lyrics but not really weird as such, opting instead to keep his concepts rooted in ghetto reality. In its name and subject matter, "Can I Borrow a Dollar" was perhaps Common's most conscious effort to capture examples of life in poor, urban America.

Handfuls of that nonfiction approach show up on Electric Circus. But so do explorations of music and lyric that will make lovers of the hardcore roll their eyes and say, damn, there goes another soldier. On "Come Close," he admits to a woman (namely, his squeeze Erykah Badu), "The pimp in me/may have to die with you." Common's thoughts may be reaching beyond the proverbial game, but those upset with him can always turn to new rappers like 50 Cent—known for his nine gunshot wounds and crack-wise experience. I think that Common is well aware of the risks he's taking, and that his exploration isn't necessarily hinged on abandoning his roots.

In a botanical sense, those roots have given life to the stems and leaves that support budding ideas on this new record. And, unquestionably, those roots are underground. But with a celebrity girlfriend and a new Coca-Cola commercial that oddly preaches "Be Real," Common is well above the surface. Hmm . . . now where did I put that 50 Cent album again . . . ?

Nevertheless, cool or weird, you can thank Common, as well as the Roots and Talib Kweli, for the guitar riffs and odes to rock n' roll that are set to litter mainstream hip-hop albums this year. Common may have bitten off more than he can chew this time around, but Electric Circus will influence the community in the near future.

💻 *Chat Room*

1. The writers of the "Chicago" arts briefs were told to narrowly angle their reviews; such a short form requires a tight focus. Read each brief and identify the angle.
2. Identify the sentence in which Nick Roskelly expresses his opinion of Common's Electric Circus. Is it thumbs up or thumbs down?

✏️ *Create*

1. Write an arts brief for a recent movie release, play or a music concert. Be sure you have an angle: a focus on a particular actor, aspect of the storyline, editing, lighting, animation, and so on, depending on what you feel is most important in the production's success or failure. Remember: thumbs up or thumbs down.
2. Write a review of a new CD by your favorite artist. As Nick does, demonstrate your knowledge of the artist's music category or the artist's previous works.
3. Write a review of a music video. Music videos combine aspects of both film and live concerts; reflect that mixture in your review.
4. Rewrite Nick's Electric Circus review as an arts brief.

℗ *View* _____

Although student writers may find ample opportunity to publish opinion pieces in campus publications, the novice professional will find few such chances. Nevertheless, some journalists eventually make careers of writing columns or reviews. Even if you don't become a professional feature writer, you may express your views anyway—on the editorial pages of magazines and newspapers where reader viewpoints and letters appear regularly.

The best opinion writing will have narrow focus, be concise and brief, and ground the writer's view in special knowledge or appropriate experience. Types of opinion pieces range from editorials to columns to arts reviews. Like all good writing, viewpoints should be expressed clearly, concisely and correctly. Writing gives voice to individual ideas, which is essential to the functioning of a democratic society.

📖 *Help* _____

columnist customarily, an experienced writer who communicates his or her opinions, usually in a specific rhetorical mode or on a narrow theme, in regular newspaper or magazine columns.

letter to the editor a letter written by a reader that is published on the editorial page of a newspaper or magazine.

op ed a full-length opinion piece written by a reader that is published on the editorial page of a newspaper or magazine. Some op eds may come from newspaper syndicates and are written by professional writers whose material is disseminated to service subscribers.

opinion piece a format that includes editorials and letters to the editor, both of which offer short, often punchy commentary on timely, controversial issues.

review an opinion piece, of varying lengths, in which a feature writer analyzes the content and quality of a book, movie, play production, dance or music concert, recording artist or art exhibit.

Notes _____

1. Robert E. Schnakenberg, "Siskel and Ebert," *St. James Encyclopedia of Popular Culture*, 26 May 2003 <http://findarticles.com/cf_0/g1epc/ bio/2419201116/pl/article.jhtml>.

2. Julie Rigg, "Pauline Kael–A Tribute," *Senses of Cinema*, 26 May 2003 <http://sensesofcinema .com/contents/01/17/kael.html>.

3. Adam Gopnik, "Barney in Paris," *Paris to the Moon* (New York: Random House, 2000) 172.

4. "Columnists—Past and Present," *Commentary & Columns*, 1 June 2003.

Bibliography

Aamidor, Abraham. *Real Feature Writing*. Mahwah: Lawrence Erlbaum, 1999.

Ad Age Almanac. 31 Dec. 2002:22.

"Ad Spending By Media." *Advertising Age* 25 Nov. 2002:10.

Advertising Age Fact Pack. 9 Sept. 2002.

Birkets, Sven. "Into the Electronic Millennium." *Language Awareness*. Ed. P. Escholz, A. Rosa, and V. Clark. 7th ed. New York: St. Martin's Press, 1997.

Bivins, Thomas. *Handbook for Public Relations Writing*. New York: McGraw Hill, 1999.

Brogan, Kathryn Struckel, ed. *Writer's Market*. Cincinnati: Writer's Digest, 2003.

Capote, Truman. *In Cold Blood*. New York: Signet, 1981.

Collins, Rives, and Pamela J. Cooper. *The Power of Story*. Boston: Allyn & Bacon, 1997.

Cuneo, Alice Z., and Tobin Elkin. "E-tailers to See Slower Holiday Growth." *Advertising Age*. 12 Nov 2001:50.

Daiker, Donald, et al. *The Writer's Options*. New York: HarperCollins, 1994.

Dimitrius, Jo-Ellan, and Mark Mazzarella. *Reading People*. New York: Ballantine Books, 1999.

Donaton, Scott, et al. "The Toughest of Times." *Advertising Age*. 22 Oct 2001: S-12.

Dunne, Dominick. *Justice: Crimes, Trials, and Punishments*. New York: Crown, 2001.

Elkin, Tobi. "AOL Music Deal: Customized Britney." *Advertising Age*. 1 Apr 2002:16.

Fast, Julius. *Body Politics*. New York: Tower, 1980.

Fine, Jon. "Magazine World Weights Its Worth." *Advertising Age*. 22 Oct 2002: S-2.

Fine, Jon. "Tongue Tied to Brand." *Advertising Age*. 17 Jun 2002:14.

Fineman, Howard. "The Brave New World of Cybertribes." *Newsweek*. 27 Feb 2001:30–33.

Flower, Linda. *Problem-Solving Strategies for Writing in College and Community*. New York: Harcourt Brace, 1998.

Franklin, Jon. *Writing for Story*. New York: Plume, 1994.

Garrison, Bruce. *Professional Feature Writing*. Hillsdale: Lawrence Erlbaum, 1989.

Giles, Carl H. *The Student Journalist and Feature Writing*. New York: Richard Rosens, 1969.

Givens, David B. *The Nonverbal Dictionary of Gestures, Signs, and Body Language Cues*. Spokane: Center for Nonverbal Studies Press, 2002. 8 Sept 2002. <http://www.members.aol.com/nonverbal2/diction1.htm>.

Goldstein, Norm, ed. *The Associated Press Stylebook and Briefing on Law Media*. Cambridge: Perseus, 2000.

Gopnik, Adam. *Paris to the Moon*. New York: Random House, 2000.

Hennessy, Brendan. *Writing Feature Articles*. 3rd ed. Oxford: Focal, 1989.

Horner, Winifred Bryant. *Rhetoric in the Classical Tradition*. New York: St. Martin's, 1988.

Hunt, Todd, and James E. Grunig. *Public Relations Techniques*. Fort Worth: Harcourt Brace, 1994.

"Interactive." *Advertising Age*. 29 Oct 2001:34.

Kennedy, George, et al. *Beyond the Inverted Pyramid*. New York: St. Martin's, 1993.

Lee, Monle, and Carla Johnson. *Principles of Advertising: A Global Perspective*. New York: Haworth, 1999.

Lichty, Tom. *Design Principles for Desktop Publishers*. Belmont: Wadsworth, 1994.

Linnett, Richard. "Magazines Pay Price of TV Recovery." *Advertising Age*. 2 Sept 2002.

Lovinger, Paul W. *American English Usage and Style*. New York: Penguin Putnam, 2000.

Mand, Adrienne. "Orbitz Stays on E-marketing Course." *Advertising Age*. 29 Oct 2001:34.

Martin, Meghan. "Journalism Panel Relates 9/11 Changes." *Observer* [Notre Dame, IN]. 17 Sept 2002:4.

Meenan, Jim. "Winter's Growing Grime Prompts Cleaning Time." *South Bend Tribune*. 5 Jan 2002:A5.

Miller, Arthur. "Introduction to the Collected Plays." *The Theatre Essays of Arthur Miller*. Ed. R. A. Martin. New York: Penguin, 1978.

Mills, Nicolaus. *The New Journalism: A Historical Anthology*. New York: McGraw-Hill, 1974.

Oates, Stephen B. *With Malice Toward None*. New York: Harper Perennial, 1994.

Pitts, Beverly J., et al. *The Process of Media Writing*. Boston: Allyn & Bacon, 1997.

Quindlen, Anna. "Weren't We All So Young Then?" *Newsweek*. 31 Dec 2001/7 Jan. 2002:112.

Rich, Carole. *Creating Online Media*. Boston: McGraw Hill, 1999.

Rivers, William R. *Free-Lancer and Staff Writer*. Belmont: Wadsworth, 1991.

Rodman, George. *Making Sense of Media*. Boston: Allyn & Bacon, 2001.

Scanlan, Chip. "Writers at Work: The Process Approach to Newswriting." *Poynter Online*. 4 Jan 2000. <http://www.poynter.org/special/tipsheets2/reporting.htm>.

Seitel, Fraser P. *The Practice of Public Relations*. 6th ed. Englewood Cliffs: Prentice-Hall, 1995.

Shakespeare, William. *Macbeth*. New York: Washington Square, 1992.

Sims, Norman, and Mark Kramer, Eds. *Literary Journalism*. New York: Ballantine, 1995.

Skidmore, Max J. *Social Security and Its Enemies*. Boulder: Westview, 1999.

"Special Report: Magazine Forecast." *Advertising Age*. 22 Oct 2001: S-1.

Steele, Bob, and Roy Peter Clark. "Coverage of the Kennedy Crash: The Crisis of Celebrity Journalism." *Poynter Online*. 3 Jan 1999. <http://www.poynter.org/special/point/Kennedy.htm>.

Teinowitz, Ira. "USA Today: No Longer a Newspaper." *Advertising Age*. 9 Sept 2002:16.

Whitaker, Richard W., Janet E. Ramsey, and Ronald D. Smith. *Media Writing*. New York: Longman, 2000.

Wilcox, Dennis L., Philip H. Ault, and Warren K. Agee. *Public Relations Strategies and Tactics*. 5th ed. New York: Longman, 1998.

Witmer, Diane F. *Spinning the Web*. New York: Longman, 2000.

Zinsser, William. *On Writing Well*. 3rd ed. New York: Harper & Row, 1985.

Index

A

Aamidor, Abraham, 62
Advertising, 120, 137–138
"After the Cheering Stops," 50–52
Alexander, Susan M.
   profile by Basinski, 125, 128–130
   query letters by, 122, 127, 130–131
   "Sports Massage Is His Passion," 125, 127
Amazon.com, 155
American Heritage Dictionary, 97
American Society of Magazine Editors, 7, 124
American Society of Newspaper Editors, 4, 142
". . . And Justice for Al," 82
Anecdote in narrative lead, 39, 56
Angle, 6, 17, 20, 38
Annual reports, 173, 180
Assignment, 19
Associated Press (AP) Stylebook, 44, 157–158
Attitudes, 96–97, 100–102, 114, 115
Attraction travel writing, 102
Attribution, 64–65, 75

B

Backgrounders, 173, 180
Basinski, Ann, 125, 128–130
Beat reporters, 140, 152
Bennet, Brian, 58, 78
Biographies, 173–174, 180
Birkerts, Sven, 154, 155, 156
Bivins, Thomas, 168, 170
Black Women, 154
Blanchard, Tara, 101, 112–113
Blocking, 63
Body, transition to, 42
Body language, 63, 98–99
Bridges, 41–42
Brown, Mary Ellen, 160–161
Brown, Tina, 120

Bureau of Labor Statistics, 169
Business features, 141, 152

C

"Cake Nazi, The," 160–161
Capote, Truman, 78
CD-ROM, portfolios on, 15
Channing, Carol, 80–81
Chekhov, Anton, 38
Chesterfield, Earl of, 36
"Chicago" (movie), 69–71
"Chicago" (review), 193–194
"Chicago: City of Cheap Thrills," 102, 109–112
Chicago Women in Publishing (CWIP) Web site, 124
Chronological transitions, 42
Citations
   attribution and plagiarism, 64–65, 75
   for Web sites, 30, 31
Clinton, William Jefferson, 182, 188–189
Closed interview questions, 80
Clothing, observing, 63, 99–100
Clustering ideas, 22
Colloquialisms, 44, 56
"Colm Feore," 46–49, 62
Color, personal style and, 44–45
Color features
   defined, 69, 152
   examples, 69–71, 144
   for newspapers, 142–143, 144, 152
Columnists, 196
Columns, 183
Colvin, Megan, 184–186
"Common's Electric Circus (MCA)," 194–195
Comparative transitions, 42
Conclusions, 43
Conducting interviews, 82–83
Context, 23–24, 41–42

Contrast
  defined, 56
  lead, 40
  transitions, 42
Coristin, Laura, 193–194
Cover story, 34
Creative briefs (direction sheet), 172–173
Credibility of magazines, 121
Curley, Tom, 138
Cut lines, 122, 136
CWIP (Chicago Women in Publishing)
    Web site, 124

D
"Death of a Salesman," 23
Descriptive lead, 37, 38, 39, 56
Destination travel writing, 102
Dimitrius, Jo-Ellan, 99
"Dining High," 144
Direct address, 40, 56
Direction sheets, 172–173
Direct observation, 62, 75
"Disaster Search and Rescue Dogs,"
    132–135
Dominello, Andrea, 5–6, 8–10, 62
Donnellon, Molly, 6–7, 10–13, 23, 62, 99
Donovan, Renée, 30–31, 45, 162–163
Dooley, Kate, 92–93
Drafts, 24. *See also* Revision
Duda, Kaitlin E.
  direction sheet by, 172–173
  "Society Challenged and Changed, A,"
    40–41
  "Tune into Reality," 77, 170, 175–176
Dunne, Dominick, 40, 81

E
Editorials, 182
"8:16 a.m.," 184–186
Electronic interviews, 77–78, 95
Emotion, feature story emphasis on, 5
Ending interviews, 83, 85
Ethics, interviews and, 84–85
Ethnic press newspapers, 139
Extended interviews, 78, 81, 84–85, 95

F
Face-to-face interviews, 77, 82–83, 95
Fact checking, 142

Feature release, 174, 180
Feature stories
  essay structure for, 37–38
  lead story, 35
  model stories in this book, 7
  news stories versus, 5
  news versus timeless, 13
  participatory feature, 7
  personal experience feature, 10
  primary feature, 115
  prize-winning, 7
  purposes fulfilled by, 5, 16
  secondary feature, 35, 115
Feature treatment of news stories, 5
Fiddler's Hearth, 162–163
First person, 46, 187
Five Ws + How, 3, 5, 16
Flashback, 56, 157
Flashforward, 56
"Flight in Bomber Evokes Wartime
    Memories," 5–6, 8–10, 62
Flower, Linda, 22, 24
Focus, shifting from self to subject, 24–25
Focus statement, 41–42
Follow-ups to query letters, 123
Food trend feature, 143, 150–151
Ford, Emily, 102, 109–112
Formal outline, 22–23, 34
Franklin, Jon, 38
"Free Falling," 144

G
Galbraith, Matthew S., 59, 66–69
Garrison, Bruce, 20–21, 102
General assignment writers, 140, 152
Generating ideas, 19–21
Giles, Carl H., 38, 39, 58, 141
Giles, Jeff, 82–83
Global revision, 24–26, 34
Gonzales, Evelyn, 32–34, 45
Group interviews, 78, 95

H
Hair style, observing, 63
Hall, Edward, 63
Handbook for Public Relations Writing, 168
Hansen, Eric, 50–52, 143
Hemingway, Ernest
  imitating, 7, 45, 53, 54

"Moveable Feast, A," 53, 54, 101
  on "why" as missing from journalism, 3, 44
Hennessy, Brendan, 18
Historical features, 141, 152
Horner, Winifred Bryant, 44
Human interest, 5–6, 38

I

"I" (first person), 46, 187
"I Am an American," 32–34
Idea generation, 19–21
Illustrations (visual aids), 122, 123
Imitation, 7, 45, 53–55
Impact statement, 38, 56
"In Cold Blood," 78
Indirect observation, 62, 75
Info box, 143, 152
Informal outline, 23, 34
In-house publications, 122
"Injustice Undone," 59, 66–69
Internet, the. *See also* Internet resources
  AP Stylebook guidelines, 157–158
  creating sites, 160
  feature writing, 157–158
  industry, 155
  interactive capabilities, 154
  newspaper industry and, 137
  online magazine home page, 159
  package story structure, 155–156
  reading habits, 155
  roundup feature, 158, 167
  shovelware, 154–155, 167
  top online national print news sites (by traffic), 156
  USA Today and, 138
  writers online, 158, 160
Internet features
  examples, 160–166
  roundup feature, 158, 167
  shovelware, 154–155, 167
  writing, 157–158
Internet resources
  for magazine writers, 124
  for newspaper writers, 142
  online reference works, 61
  prize-winning feature stories, 7
  researching, 61
  search engines, 75
  speeches, 171

"Internship Payoff, The," 15–16
Interpretation, 97–100, 115
Interviews. *See also* Personal observation; Personality profiles
  conducting, 82–83
  electronic, 77–78, 95
  ending, 83, 85
  ethics and, 84–85
  extended, 78, 81, 84–85, 95
  face to face, 77, 82–83, 95
  group, 78, 95
  overview, 94
  preparing for, 78–80
  purpose of, 76–77
  of sports figures, 143
  telephone, 77, 83, 95
  time factor for, 80–81, 83
Inverted pyramid structure, 4, 17, 37

J

James, Holly M.
  "Sean Savage," 62, 71–73
  "Sweet Onion Charm," 97, 101, 102, 105–107, 141
"James Earl Jones," 85–88
Jargon, 44, 56
Jobvine Web site, 124
"John of the Archives," 170, 177–178
Johnson, Carla
  "Colm Feore," 46–49, 62
  "James Earl Jones," 85–88
  "Judeo Jerry," 183, 191
  "Leader Much Like Ourselves, A," 182, 188–189
  "Marin Mazzie," 100, 102–103
  "Marz Sweet Shop," 141, 146–149
  "'Rent' Reclaims Broadway for U.S.," 104–105
  "Roxie Hart," 69–71
  "Some Like It Hot," 192–193
  "Troy Donahue," 27–29
Jones, Jennifer, 7, 45, 53–55
Jonusas, Kristina V., 123, 171, 172
"Judeo Jerry," 183, 191

K

Kael, Pauline, 181
Kelley, David E., 6
Kennedy, Jr., John F., media criticized after death of, 4

Kirkbride, Jennika, 144–146
Kovach, John, 177–178
Kramer, Mark, 46, 137, 138
Krieger, Liz, 39

L
Larson, Jonathan, 100, 101, 102
Lead
    angle and, 38
    contrast, 40
    defined, 4, 17
    descriptive, 37, 38, 39
    direct address, 40
    impact statement, 38
    in inverted pyramid structure, 4
    for inverted pyramid structure, 37
    narration (anecdote), 39
    of news stories versus features, 37–38
    question, 41
    quotation, 40–41
    revising to shift focus, 24–25
    suspended-interest, 3, 38
"Leader Much Like Ourselves, A," 182,
        188–189
Leading questions for interviews, 79
Lead story, 35
Lehmann, Nicholas, 97–98
Length
    of Internet features, 154, 156
    of magazine features, 122
    of query letters, 122, 123
    of reviews, 183
Letters to the editor, 182, 191, 196
Library resources, 59–61, 75
Lifestyle features, 143, 152
Linear structure, 156–157, 167
Linked features. *See* Internet features
"Literary Journalism," 46, 137
Loaded interview questions, 80
Localisms, 44, 56
Local revision, 24
Local tie, 20, 35
"Looking to Get Hitched, Just Not Yet,"
        189–190

M
Magazine Publishers of America Web site,
        124
Magazines
    fact checking, 142

feature market, 122–125
features, 121–122
first person and, 187
in-house publications, 122
newspapers versus, 139–140
niche magazines, 120, 136
overview, 119–121, 136
query letters, 122, 123, 127, 130–131,
        136
resources for writers, 124
top U.S. magazines (by circulation),
        120
Magness, Sarah K., 170, 177–178
"Making Stone Soup," 6–7, 10–13, 23, 62,
        99
"Marin Mazzie," 100, 102–103
"Marz Sweet Shop," 141, 146–149
Mazzarella, Mark, 99
Mazzie, Marin, 100, 102–103
McCain, John, 97–98
Medill School of Journalism, 142
Mehus-Roe, Kristin, 132–135
"Men's Pickup Lines, Then and Now,"
        30–31
Michalski, Amelia, 108
Miller, Arthur, 23
Moor, Bill, 3–4, 38
Morgues, 59
"Moveable Feast, A," 53, 54, 101
"Moveable Moscow, A," 7, 45, 53–55

N
Narrative bridges, 42
Narrative lead, 39, 56
National Magazine Awards, 7, 10
Naughton, Jim, 4–5
Neuharth, Al, 138
"Newest Member of the Footwear Family,
        The," 144–146
"New Journalism," 46
News features
    defined, 17
    emergence of, 5
    timeless features versus, 13
Newsletters, 171, 180
Newspaper features
    business features, 141, 146–149, 152
    categories, 140
    color features, 142–143, 144, 152

examples, 144–151
fact checking, 142
historical features, 141, 146–149, 152
lifestyle features, 143, 150–151, 152
magazine features versus, 139–140
overview, 139–140
resources for writers, 142
sports features, 143, 152
trend features, 141, 144–151, 152
Newspapers
classifications, 138–139
geographic targeting by, 138
magazines versus, 139–140
overview, 137–139, 152
presses, 139
sizes, 139
top U.S. newspapers (by circulation), 138
writers and reporters, 139–141, 152
News releases, 180
News stories
changing style for, 3, 4–5
feature stories versus, 5
five Ws + How for, 3, 16
inverted pyramid structure for, 4, 37
news features, 5, 13
purpose of, 37
The New York Times online, 154–155
Niche magazines, 120, 136
"No. 1 Reason to Visit South Bend—The Cake Nazi," 160–161
"No. 2 Reason to Visit South Bend—There's No Right Way," 161–162
"No. 10 Reason to Visit South Bend—You Can Fiddle Away at Fiddler's," 162–163
Nonlinear structure, 156–157, 167
Nut graph or nut paragraph, 41–42, 56

O
Online magazine home page, 159
Online package, 155–156, 167
On Writing Well, 18
Op eds, 182, 196
Open-ended interview questions, 79
Opinion pieces. *See also* Reviews
columns, 183
defined, 196
editorials, 182
examples, 184, 184–195
letters to the editor, 182, 196

op eds, 182, 196
overview, 182–183
for public relations, 174, 180
Orfanos, Marianne, 193–194
Organization of material. *See also* Structure
attitudes and, 100
clustering ideas, 22
outlines, 22–23, 26, 34, 35
overview, 21–23
Outlines, 22–23, 26, 34, 35

P
Packages online, 155–156, 167
Package story structure, 155–156
Participatory features, 7, 17
Pendley, Sara, 30–31, 45, 161–162
Personal experience features, 10
Personal experience travel writing, 102
Personality profiles. *See also* Interviews
avoiding "I" and "you" in, 46
defined, 35
examples, 46–52, 71–73, 125, 127
Personal observation
body language, 62, 98–99
direct versus indirect, 62, 75
environment, 99
interpretation and, 97–100
in interviews, 76–77, 82
overview, 61–62
personal appearance, 63, 99–100
props, 63
proxemics, 63, 99
voices, 62, 98
Personal style. *See also* Style
color and tone, 44–45
feature story emphasis on, 5, 6
imitating, 7
overview, 44–46
Persuasive platforms (direction sheets), 172–173
Photo gallery, 167
Photographs (visual aids), 122, 123
Pixel, 156, 167
Plagiarism, 64–65, 75
Portfolio, 14–15
Preparing for interviews, 78–80
Presses for newspapers, 139
Press releases, 170–171, 180
Previews, 60–61, 75
Primary feature, 115

Private libraries, 75
"Problem-Solving Strategies for Writing in College and Community," 22
Process of feature writing
   context, 23–24
   drafting and revision, 24–26
   idea generation, 19–21
   organization, 21–23
   overview, 34
   as problem solving, 18–19
   research, 21
"Professional Feature Writing," 20–21
Profiles, 19, 35. *See also* Personality profiles
Pronouns, 46, 187
Proposals, public relations, 170
Props, observing, 63
Proxemics, 63, 99
Publication style, 44
Public libraries, 59–60
Public records, 59, 75
Public relations
   defined, 180
   employment opportunities, 168–169
   formats, 168–169
   speechwriting, 171
   writers, 180
   writing, 180
Public relations features
   direction sheet, 172–173
   examples, 175–178
   overview, 179
   proposals, 170
   purpose of, 169–170
   release format, 174
   researching, 170
   types of, 170–171, 173–174
Pulitzer Prize awards for features, 7

Q
Query letters
   crafting, 122, 123
   defined, 136
   examples, 127, 130–131
   follow-up letters, 123
Question lead, 41, 56
Quindlen, Anna, 39
Quotation lead, 40–41, 56
Quotations, 76–77, 100–101

R
Radio and Television News Directors Association, 4
Reader, keeping foremost, 18
Reading, 7, 155
"Reading People," 99
"Real Feature Writing," 62
Rejection, handling, 124–125
"Rent" (Larson rock musical), 100, 102–105
"'Rent' Reclaims Broadway for U.S.," 104–105
Research
   importance of, 58
   Internet resources, 61
   for interviews, 81
   library resources, 59–61
   multiple sources for, 58
   overview, 21, 74
   personal observation, 61–63
   public records, 59
   for public relations features, 170
   time required for, 59
   verifying facts, 61
Reviews
   defined, 196
   examples, 192–195
   first person and, 187
   overview, 183–184
   successful reviewers, 181–182
Revision
   defined, 24, 35
   examples, 24–26
   global, 24, 25–26, 34
   local, 24
   for online publication, 156, 157
Rhetorical modes, 42–43, 56
Roskelly, Nick, 158, 160, 164–166, 194–195
Roundup features, 158, 167
Roundup travel writing, 102
"Roxie Hart," 69–71

S
Sanchez, Sonia, 154
Scanlan, Chip, 5
"Sean Savage," 62, 71–73
Search engines, 75
Secondary features, 35, 115
Second person, 46
Service travel writing, 102
Shakespeare, William, 96

Sheridan, Richard Brinsley, 119
Shovelware, 154–155, 156, 167
Sidebars. *See also* Color features
   defined, 100, 115
   example, 108
   info boxes, 143, 152
   for magazine features, 124
   overview, 100
"Silent Language, The," 63
Sims, Norman, 46
Sizes of newspapers, 139
Skidmore, Max J., 63, 64–65
Slang, 44, 56
"Slice of Comfort, A," 143, 150–151
"Smiling in Terror," 164–166
"Society Challenged and Changed, A," 40–41
Society of Professional Journalists (SPJ), 7, 84
"Some Like It Hot," 192–193
"Something Old, Something New," 92–93
Special assignment writers, 140, 152
Speechwriting, public relations, 171
Sports features, 143, 152
"Sports Massage Is His Passion," 125, 127
Stahl, Brandon, 142
Static Magazine, 164–166
Stories, 5, 16, 17
Structure. *See also* Lead; Organization of material
   body paragraphs, 41–43
   characteristics, 44
   chronology and, 42–43
   conclusion, 43
   feature leads, 38–41
   Internet package, 155–156
   linear versus nonlinear, 156–157, 167
   for magazine features, 121–122
   nut graph, 41–42
   organization of material, 42–43
   overview, 55–56
   for reviews, 183–184
   rhetorical modes, 42–43
   style versus, 36–37
   transition to body, 42
Strzelecki, Molly, 40, 143, 150–151, 189–190
"Student Journalist and Feature Writing, The," 58
Style. *See also* Personal style
   levels of formality in, 36
   for magazine features, 121
   for news features, 140
   for opinion pieces, 182
   personal style, 5, 6, 7, 44–46

publication style, 44
   for public relations writing, 169–170
   for reviews, 184
   structure versus, 36–37
"Sum of Two Evils, The," 58, 78
Suspended-interest lead, 3, 38
"Sweet Onion Charm," 97, 101, 102, 105–107, 141
Symbols concluding newspaper copy, 174

T
Tabloids, inaccuracies in, 61
Teasers, 154, 167
Telephone interviews, 77, 83, 95
"There's No Right Way," 161–162
Titles for online stories, 157
Tone
   attitude and, 97, 101
   defined, 97, 115
   for newspaper features, 140
   personal style and, 44–45
Transitions, 42, 44, 56
"Trash, Art and the Movies," 181
Travel features
   attitude in, 101–102
   defined, 101, 115
   examples, 105–113
   types of, 102
Trend features
   defined, 35, 175
   examples, 144–149, 175–176
   finding topics for, 20
   newspaper features, 141, 144–151, 152
   for public relations, 175–176
"Troy Donahue"
   clustering for story, 22
   feature story, 27–29
   outline for story, 26
   personal observation for story, 63
   process of feature writing, 18–19
"Tune into Reality," 77, 170, 175–176
"Two Beaches," 101, 112–113

U
United Press International (UPI) stylebook, 44
Unity, 44, 56
USA Today, 138

V
Vanity Fair, 183

Verifying facts, 61
"Vidalia Onion Festival," 108
Visual aids, 122, 123
Voelkel, Theodore S., 157
Voices, observing, 63, 98

W
Web resources. *See* Internet resources
Web sites, citing, 30, 31
Weisskopf, Michael, 58, 78
Williams, Nellie, 15–16, 24–26
Wolff, Michael, 121
Women in Publishing Boston Web site, 124
Women in Publishing Web site, 124
"Working Dogs," 132

"Working Man: Tom Wopat," 89–92
Writer's Market, 124, 136
"Writer's Words, A," 124, 128–130
Writing portfolio, 14–15

Y
Yearbook, college, 170
Yes/no interview questions, 79
"You" (second person), 46
"You Can Fiddle Away at Fiddler's,"
    162–163

Z
Zinsser, William, 18